I0424424

Thought Genesis

Thought Genesis

◆

The Evolution of Reason

David M. Sternberg

iUniverse, Inc.
New York Bloomington Shanghai

Thought Genesis
The Evolution of Reason

Copyright © 2008 by David M. Snir

All rights reserved. No part of this book may be used or reproduced by any means, graphic, electronic, or mechanical, including photocopying, recording, taping or by any information storage retrieval system without the written permission of the publisher except in the case of brief quotations embodied in critical articles and reviews.

iUniverse books may be ordered through booksellers or by contacting:

iUniverse
1663 Liberty Drive
Bloomington, IN 47403
www.iuniverse.com
1-800-Authors (1-800-288-4677)

Because of the dynamic nature of the Internet, any Web addresses or links contained in this book may have changed since publication and may no longer be valid.

The views expressed in this work are solely those of the author and do not necessarily reflect the views of the publisher, and the publisher hereby disclaims any responsibility for them.

ISBN: 978-0-595-50133-5 (pbk)
ISBN: 978-0-595-49677-8 (cloth)
ISBN: 978-0-595-61386-1 (ebk)

Printed in the United States of America

For Mother, who has given me life long after birth.

For Father, who has always been the source of my inspiration.

For my two loving sisters, who have always been there
for support and encouragement.

And for my twin brother, who I am not complete without.

All humanity could share a common insanity and be immersed in a common illusion while living in a common chaos. That can't be disproved, but we have no choice but to follow our senses.

—Isaac Asimov, *Foundation's Edge*

Contents

Introduction

Through countless ontological efforts, the human race has considered itself the center of all things. This conceit has changed piecemeal since the Enlightenment, a movement that inquired into the workings of the natural world based solely on reason and evidence, without regard for religious beliefs.

Since Darwin's ingenious observations, the superciliousness of humankind has diminished to proportions considered unfavorable by some groups. These groups believe only one creation of God has a soul and only one creation was gifted with the ability to think and to rule all others. Following the same methods of validation, scientists and thinkers used to promote these theories using evidence. These groups found it very easy to provide evidence that humans were unlike any other animal. Something must have presented humans with their exceptional minds. Nonhumans did not build beautiful castles, write literature, speak, or conquer horizons—all that begins with human thought, as pure, or as ingenious, or even as rogue as it might be, and ends with its execution, which is possible only by the human hands and voice.

My goal in writing this book is to harmoniously unite several disciplines of research—some are centuries old and some are drawn directly from recent ongoing research in the study of the mind—into one coherent and unified work that will address the origins of our talented mind.

Although this work might be rendered as a mechanistic view of the human mind, I am convinced that through careful reading the reader will reach the same conclusions to the origins of our thought as I have. In that respect, I have tried to compose a philosophical approach to the subject matter but have drawn my conclusions from physiological evolutionary standpoints and recent research in cognitive neuroscience. Cognitive neuroscience is a multidisciplinary field of research that combines the study of behavior, neural networks linguistics, biology, and even artificial intelligence to understand how animals react to their surroundings.

Throughout this book, I will bring references from both the biological and technological worlds. Some serve merely for illustrative purposes, but others are needed to establish the correct conclusion and to understand how vast the scope of its application can be.

UNIT I

Chapter I: A Very Brief History of Evolution

In early stages of materializing this book, it became obvious that before discussing the evolution of logic we must first discuss the evolution of our eco-system. In this universe, everything is built with some measure of order (which we may or may not completely understand) but especially with reoccurring patterns of some sort. Later in this book, the connection between prehistoric events and animal behavior to our modern everyday complex social networking will become irresistibly obvious. However, we must first have a fair ground upon which we should build our discussion.

This book will not be centered toward human behaviors alone. To answer why we think the way we do—or to be more exact, why we act the way we do—it is important to ask first when and how we began to act the way we act. All species, including humans, are affected by environment, the one thing that has existed since Earth's first steps toward a planet. The Earth's environment is not a constant, fixed environment. As the planet travels through space orbiting the sun, many factors affect its environment, such as the tilt of its axis. As the environment changes, life-forms adapt, thrive, or become extinct. Although extraterrestrial visits of rocks can eliminate an entire crop of living organisms on a planet, natural environmental changes can also inflict damage, sometimes with catastrophic results. A good example of an extreme natural environment change is our modern ice age, three million years ago, when glaciers froze the water supply. Continents were either linked or separated from each other because of lower sea levels, which caused the migration—and in many cases, the elimination—of animals. However, environmental alteration, such as a global climate change, can also promote birth to life-forms. A long cooling period, from around fifty million years ago (Eocene) to about ten thousand years ago (Pleistocene), made it possible for living organisms to exist, and the warming of Africa, which was prehistorically a thriving forest, caused our ape ancestors to leave their now-disappearing trees and begin the story of our bipedal ancestors.[1],[2]

Human Evolution

The theory of evolution has always been the subject of much debate. I am not referring to the creationistic ongoing battle against reason, but to the enduring debate over the origin of the first man-ape ancestors, their bipedalism, brain-to-head ratio, head-to-body ratio, and many other factors that are on the front line of fire. Paleontology, the study of fossils, relies on two main data streams: biology and geology. Paleontologists believe the environment is a key factor in completing the hominid species' timeline. One of the earliest identified hominid species was *Australopithecus ramidus*. Although classified as an ape and not a human species, this four to five million-year-old ape ancestor, which lived in northeast Africa in the Hadar region of Ethiopia, was found by the hip joint and the pelvis to be an erect, bipedal species.[3] Contrary to previous assumptions, an intriguing argument regarding bipeds suggests that early hominids walked on two legs prior to the disappearance of forests in eastern Africa.[4]

The climate change resulted in the disappearance of forests in eastern Africa and played an important role in the development of our brain. Not only did the climate change force us away from our comfortable trees, pushing us to exercise our bipedal potential, but it also took away vast source of plantation, which served as our main source of food. As a result, our appetite for flesh grew stronger. Meat, a protein-rich supplement to vegetables, was rocket fuel for our brains. In that period, we started growing larger and larger brains, and it is almost surprising to see how fast our brain did grow, considering how evolution worked so far.

During this time the prehistoric mammalian species developed the neocortex. The neocortex is perhaps the greatest gift higher-functioning beings possess. The neocortex is an extra layer of brain that all mammals have in common. Located at the top of the cerebral hemispheres, the six-layered gray matter surrounding the white matter of the cerebrum is responsible for our innovative developments and our highly developed language. Although involved in all our higher functions, such as sensory perception and motor commands, scientists debate to what extent the neocortex affects or is responsible for our cognitive processes. It is clear that many other animals that lack this part of the brain exercise cognitive processing without this idiosyncratic, six-layered neocortical formation.[5]

Because we were lacking sharp teeth, claws, and the other traits our worthy adversaries had, the incredible handgrip that we inherited from our tree-resident ancestors was our only defense. It is probably through chance, together with our extraordinary hands and our newly developed cortex, that we learned how to make tools: tools to slice, tools to break and crack bones, and tools to make other type of tools. The better tools we made to fight and to extract nutrients from our food, the larger our brains grew.[6]

Chapter II: The Primordial Soup—Microcosmic and Autopoietic Civilizations

Before we were humans, before we were our pre-human ancestors, and long before anything existed that was remotely close to what we define as an animal, we were a single living cell. There are many theories to how microscopic cells were originally created. Some say another planetary residue, also known as an asteroid rich with frozen water, brought them here. These theories, such as Panspermia and other exogenic ideas are rooted from the writings of the Anaxagoras, a Greek philosopher of the fifth century B.C.E[7] and other pre-Darwinian thinkers such as Benôit de Maillet (1656-1738) who speculated how sea creatures developed into land forms over time.[8] Although these theories are fascinating, they only displace the possibility of spontaneous generation and do not solve the mystery itself.

A most difficult task we can face is to find out how life started on the planet. The creation myth in the book of Genesis states that God created man from dust. The book of Genesis therefore states that a living creature was created from a non-living material, a notion that at its core corresponds to scientific thought. If we are to support the well researched and well accepted theories of the evolution of our universe we must find, or give plausible speculations to how non-living materials paved the way to the creation of living cells whether they were created here on earth or on other planets.

This theory of creation must not be concluded from unsupported evidence and must certainly not include any notions that are supported by means that reject our laws of physics and our understanding of the natural world. If we are to obtain reliable evidence, they must come with a reasonable degree of certainty pertinent to the knowledge of the evolution process.

Through countless observations of the genetic code which is universal to all life, we are certain that all life on earth is related. This necessitates that there had to be one last universal common ancestor (often referred to as LUCA).[9] The identification process of such an ancestor often takes place in sedimentary and extreme environments such as hydrothermal vents or volcanic rock that until recently were considered inhospitable for life. These microenvironments are suitable candidates for this research since they were the norm in early earth.[10]

The resolution for the creation of life should not be concluded from microfossil studies of such an ancestor alone but must also be drawn from findings of chemical evolution, a process by which the organic molecules that formed the first unites of life were created. It is important to notice that chemical evolution does not stray from the theory of natural evolution and indeed coincides to its core notion of natural selection.[11]

A growing field of research, Abiogenesis study tries to answer several of the questions that are presented when confronting these complex matters. The general consensus in Abiogenesis is that life emerged from early earth more than 3.8 billion years ago. In perspective, the earth is considered to be 4.5 billion years old and the first ocean had condensed 4.4 billion years ago.[12] There are many different theories of how non-life materials paved the way to life; however, they are all derived from the idea of spontaneous creation. This is by no means the classical notion of spontaneous creation which held that living organisms are generated from decaying organic substances, e.g. the spontaneous appearance of maggots in meat.[13] These theories of spontaneous creation vary from each other by which they proceed, but all their foundations are built on the same research.

Other than the possibility of delivery by extraterrestrial objects, a theory of organic synthesis by energy resources such as electrical discharges or ultraviolet light and a theory of organic synthesis driven by impact shocks have both been proposed. Experiments have shown that in both sources for organic synthesis, spontaneous creation can occur.[14]

An example for such an experiment is the famous Miller-Urey experiment that took place in 1953. Stanley Miller, a graduate student at that time together with his professor, Harold Urey, conducted an experiment that proved organic molecules could have spontaneously formed from inorganic materials at the conditions that were present in early earth. The experiment tested the Oparin and Haldane's hypothesis which proposed that the environment on early earth, together with chemical reactions synthesized organic compounds from inorganic precursors.[15]

Using a mixture of water (H_2O), methane (CH_4), ammonia (NH_3) and hydrogen (H_2) gases all sealed but interconnected to each other in sterile tubes. They simulated the conditions on early earth by heating up water to simulate water evaporation. This led water vapour to another flask containing a pair of electrodes that sparked to simulate lightning. The atmosphere was then cooled again so that water could condense and drip back to the first flask to be heated again in a continuous cycle.

After a week of continuous operation Miller and Urey observed that the experiment yielded surprising amount of organic compounds. Amino acids which are essential to the formation of proteins in living cells were present together with other organic compounds such as sugars and lipids that are the building blocks for nucleic acids such as deoxyribonucleic acid (DNA) and ribonucleic acid (RNA).[16]

The Miller-Urey experiment showed the possibility of organic synthesis from non organic. Many more experiments that followed were inspired from their experiment and proved to be as successful as the first. Although these organic compounds are far away from a self replicating life-form, it is very likely that their accumulation provided an adequate ground from which chemical evolution might have taken place.

This leads us to the question of self-organisation and more importantly self-replication. The organic compounds must have somehow increased in complexity. Self-organization could have either occurred by attraction or repulsion without being guided or managed by an outside source, and eventually becoming an autocatalytic set that is able to replicate itself.[17] It is important to notice that self-organisation can be observed in biological systems from the sub cellular to the large ecosystem level. In that sense, this could be rendered as a question that is related to the emergence theory which refers to the way complex systems and complex patterns arise out of a collective of simple interactions.[18] In later chapters the notion of a collective of simple parts forming complex patterns will take a much larger role in correlation to the origin and evolution of thought. However, for now it should be mentioned that it is not only an integral part in organic evolution but can also be observed occurring in many other forms and fields of sciences such as physics, chemistry, mathematics and computer science and even economics.

A cell, the basic unit of life from which all organisms are made, is a magnificent masterpiece of perfectly balanced and orchestrated members. The cell is a collection of different mechanisms working together. This book emphasizes nature's continuous use of patterns, and organisms are no exception. Most known elements in our universe are made of groups of things working in conjunction. These groups form another unit that serves its purpose as a part of another group to form another unit and so on. When these units work together, they work as a part of a microscopic organized civilization. Each part has its rules and a role in serving some greater role. I am not suggesting they have understanding to what they are creating, but same can be said on our own civilization. Numerous types of cells exist; some function as organism unto themselves and some in groups that make up a larger organism. They all share common structures and all have a nucleus (in which the DNA is stored) and different organelles to carry the vital operations of the cell.[19]

Cells tend to group together, and scientists have been using the process of cell growth and grouping in artificial cultivation in petri dishes. A petri dish simulates nature by giving a cell a place to grow and feed. The plate is usually filled with agar, along with a mixture of nutrients, salts, and amino acids.

Petri dishes have numerous uses in the field of microbiology but can also show us from where we started. We do not have to look to nature to see an organized system of cooperative individual parts. An example is the ever-expanding field of technology, in which humans have mimicked the works of nature in electronics. Before the introduction of multiprocessors or central processing units, which are often referred to as the brains of electronic machines, single units called transistors made calculations possible when grouped together with other transistors in an electrical circuit. The transistor, just like our own cellular member, reacts as energy passes through it. One of the most important inventions of the twentieth century, the transistor is a small electronic device that contains a semiconductor and has at least three electrical contacts. The transistor is the fundamental building block of computer circuitry.

Without getting into too much detail of how this key element works, a simplified look at it unveils a semiconductor material, such as silicon, that can change its electrical state when voltage is applied to it. The three basic electrical contacts are the gate, the source, and the drain.

In the race of faster computing, we combine more and more transistors to satisfy the growing demands of our electronic machines. It is important to note that this would not be possible without also minimizing the size of these transistors. Computers from the sixties were equipped with a processor built from three transistors tied together in a half an inch square space, where today, the same amount of space holds millions of miniscule transistors.[20]

Without making the transistors smaller, a computer the size of the Empire State Building would be required to perform the same calculations that my palm pilot can. Everything must be in proportion to us, just like our cells.

This is where logic can be seen. Now, I am not referring to complex logic calculations but it is worth mentioning that neither computers nor living organisms will function if the rather "simple" operations do not cooperate for some tasks.

Each part has a role, and, when it is added to a group, further operations can be obtained.

I would like to illustrate this further. Let's examine a group of three children who are engaged in a game. Each child has only one reactive operation he or she can perform. Child A can only pass the ball to her right when a ball passes by her, and Child B can only shoot to the basket when a ball pass on his left (with fail-safe accuracy). We can add many more players to this game. For example, we might add an individual who can steal the ball from any player and pass it to someone else or an individual whose sole purpose is to protect the ball from being taken by another individual. These children serve as a primitive example of team-work for winning a game. Although it might be amusing to take out a player, the goal of the game would not be reached, especially when the individuals cannot do anything further to win the game because of their limited reaction formula.

The team has a minimal purpose, and the team is built from a group of opera-tors who all have their own simple purpose, yet this does not stop here. We can investigate the individual operators and find out that they are actually bundles of smaller operators. Consider that the reaction of each operator in our team is built from many more operations, such as sensing, grabbing, passing, raising hands, and so on.

A groundbreaking research study conducted in the sixties by Lynn Margulis brought the world of science one step closer to solving the mystery of life. She focused her research on mitochondria, cells that produce energy in all high-level organisms and are members of each cell in our bodies, and she found that mito-chondria have their own genes.

This discovery led to the suggestion that at some point in history mitochon-dria were bacteria that existed in their own right. Although it is not known when the cell somehow took residence in a larger cell, it is certain that mitochondria would have benefited from protection under this larger cell. In this symbiotic relationship, mitochondria cells gave off excess energy. From their joint venture came a new cell that would eventually become the building block for all complex life-forms.[21] To sum it up, we are all high-level creatures that are—in their own right—communities of autonomous creatures that have integrated into each other.

I shall discuss in later chapters the relationship between the micro-bacterial soup to the macro world and the interaction between species. After all, this is what this book is all about. For now, I would only like to convey that sometimes these biological seeds do not do exactly what they are built to do. These operators misbehave where their genetic code has been altered, mostly because of malfunctioning at the reproduction or replication stages or because of environmental factors; sometimes with incredibly amazing or catastrophic results, both are important to the process of evolution. Environmental demands impose some form of obstacle or challenge to the living organism. Operators that adjust to these demands will prevail. A mistake or interference in genetic code will be a mutation at first, but it will re-accrue many more times as the living organism reproduces. Should the mutation result in an improvement for the organism to deal with its surroundings, it may continue as the surviving form of that code.

In the electronic world, we change our electronics more often than some would like. USB thumb drives with 2 MB flash memory are now used as door holders because today's market demands a 2 GB. Other than for archival purposes, I do not see anyone using a 368 to perform anymore. They are extinct.

So we share much more in common with our human-created machines than it may seem. It appears that everything in this world exists through some form of a common pattern, and, even if it is not the case, it seems that this is our only way to identify how things work around us. So how do we perceive the world around us?

UNIT II

Chapter III: From Cellular Processing Units to Central Processing Units (CPU)

Lex III: Actioni contrariam semper et æqualem esse reactionem: sive corporum duo- rum actiones in se mutuo semper esse æquales et in partes contrarias dirigi.

"All forces occur in pairs, and these two forces are equal in magnitude and oppo- site in direction."

Newton first published his laws of motion in his work *Philosophiae Naturalis Principia Mathematica* (1687). These laws formed the basis for classical mechan- ics.[22]

Simplifying his second law, we can extract that all action has a reaction. Just like Newton's laws of emotion, many processing units, including our brain, react to an action that was acted upon them.

From the simplest microscopic organism to the most complex life-form, the key to survival is awareness of surroundings. We are all equipped with sensory systems that are essential to our survival. Vision, hearing, touch, taste, and smell are necessary to the survival of every animal. Some animals are more dependent on one sensory system than others are. Environment through evolution advances the skills of one sensory system or another, depending on which senses are most needed. Humans were not equipped with the intense sense of smell that other mammals have. Our vision, although unique, is no match to some birds. We do not possess overdeveloped senses that are more advanced than other animals, but still we manage to adapt to our surroundings more than any other animal on this planet. Why were we lucky not to posses these superpowers? The answer is sim- ple. Through evolution we developed extremely sophisticated methods to over- come our physiological limitations, and, through the combination of our sensory systems and our developing brains, we became what we are today. So what are these sensory instruments? How do we animals use them?

To understand any sensorial system—not just in the field of neuro-science—we must understand that all a sensor does is transmit impulses from one end to the other. A dedicated cell member is assigned for the job. All organisms have at least one simple sensory cell. When the cell is stimulated, it sends impulses that are then identified by a processing unit. I will go into more detail later on how these impulses are being identified through patterns of these impulses. Humans share with other high-level animals a complex network of these neural networks, which are also called nervous systems.[23]

The Nervous System at a Closer Look

Nervous tissue is composed of neurons and glial cells. Think of an electrical wire. It has the metal core that transmits the electrical current, and a nonconducting material, such as plastic, protects this metallic wire. Similar to common electrical wire, the neurons transmit nerve messages, while glial cells protect the neurons. Since the world of science discovered this amazing property, many advances in technology have become available to not only detect these messages but also to recreate them and send them to appropriate places in test subjects, giving neuro-scientists an advantage in demystifying the works of the brain system. Through many years of numerous achievements in the world of chemistry and technology, we have all the tools necessary to understand how an organic element, such as the nerve, can create and read electrical messages.

There are about 100 billion neurons in the human brain alone. All have three parts: dendrite, cell body, and axon. The dendrites are short, branched fibers that carry signals that they receive from another cell to the cell body. The cell body will then conduct messages away through longer fibers called axons.[24]

Neuron Structure Illustration

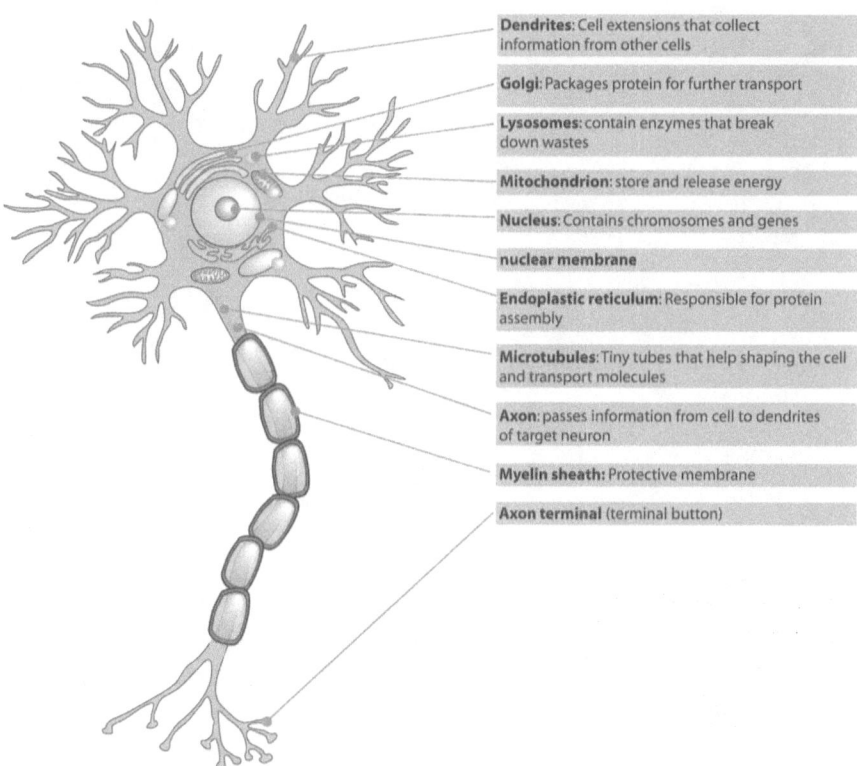

Dendrites: Cell extensions that collect information from other cells

Golgi: Packages protein for further transport

Lysosomes: contain enzymes that break down wastes

Mitochondrion: store and release energy

Nucleus: Contains chromosomes and genes

nuclear membrane

Endoplastic reticulum: Responsible for protein assembly

Microtubules: Tiny tubes that help shaping the cell and transport molecules

Axon: passes information from cell to dendrites of target neuron

Myelin sheath: Protective membrane

Axon terminal (terminal button)

Before we discuss what a nerve message is we should introduce yet another member of our neural network is the synapse. The synapse is a junction between a nerve cell and another cell in which a nerve impulse passes from an axon to the conjoined cell. The message (nerve impulse) that travels from the axon to the cell has to jump through the synaptic cleft, a space between the two cells, which requires the release of chemicals from the end of the axon. These chemical neurotransmitters are stored in tiny synaptic vesicles and are released when a nerve impulse arrives.[25]

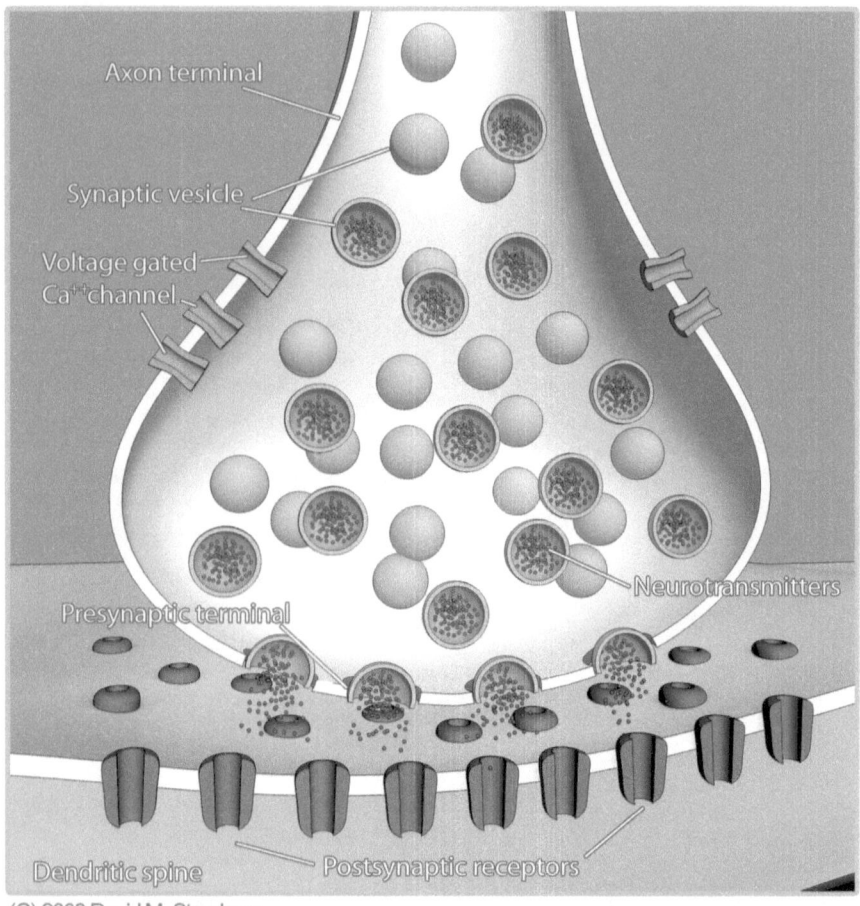

Synapses Structure Illustration

Axon terminal

Synaptic vesicle

Voltage gated
Ca^{++}channel

Neurotransmitters

Presynaptic terminal

Dendritic spine — Postsynaptic receptors

(C) 2008 David M. Sternberg

Before we go into further detail of the neural networks, we should discuss what runs through these complex networks. These networks communicate to each other, but what is being transmitted when they talk? The answer is rooted in the model of the electrical wire, which was discussed in the beginning of this section. Our nervous system communicates through electrical impulses, similar to an electrical wire.[26] The process through which all neurons communicate is chemical energy. A changed polarity of the nerve membrane, which is a bit more complex than an on\off electrical pulse, results in propagation of impulses along the membrane.

In what is also called action potential, the polarity of the membrane changes when positively charged sodium ions move inside the cell body. The membrane, which in its resting position is less positive inside than outside, is now more positive inside than outside. Now that the polarity of the membrane is opposite, potassium ions flow out of the cell, restoring the resting potential net charges. Later, sodium ions are pumped out of the cell, and potassium ions are pumped into the cell, which restores the original polarity and results in the original distribution of ions.[27]

Electric pulses have been serving us throughout history, not only as the driving force of our functioning brain, but also in the evolution of our manmade technology. The question arises, how does a series of pulses function as a way of communication? The Morse code is one example. Invented by Samuel Morse and Alfred Vail and refined by Friedrich Gerke, the international code system is represented by the duration of single tones and spelled by a series of dots, dashes, and spaces. First used with the signaling of flashlights, the code was later transmitted over great distances through electrical pulses. The sender would transmit a message over an electrical cable, which would then be interpreted.[28] For example, a person would send a telegram to inform his or her arrival to London. The original message would read, "arrive in London." The message sent would be "._. ._. ._. ..... _ . .. _. ._.. _ _ _ _. _.. _ _ _ _."

When the connection was received, an interpreter would decode the message and write it in English. Our brains are capable of recognizing an unimaginable amount of patterns that are assembled through a series of impulses. These patterns are assembled to create other high-level patterns, which are eventually categorized and fed to and from the inner communication systems of our body.[29] I will soon discuss one of the most important parts of our working brain, our memory that is responsible for "knowing" these patterns or creating and storing new ones.

Before we venture off into the workings of the brain, I would like to mention the coexistence of manmade pulsating electrical devices and natural ones. Many science fiction writers envision a future world in which the next big development in human evolution is the cyborg-sapien human, who is aided or controlled by the melding of mechanical and electronic devices. Although that future sounds far, it is already the world of today. Medicine has used machines to keep patients alive for quite some time now. There are kidney dialysis units, pacemakers for hearts, and even chips that help in connecting damaged connections between nerves. Further developments in the field of neurotechnology have successfully tackled problems that seemed unsolvable: controlling mechanical hands and legs through brain signals and bringing sight to those who are unable to use their natural eyes. Although, this infusion of electrical and biological elements is still in its first steps, the future of the cyborg is real.[30]

Now that we have touched on the workings of the nerve system, we can zoom out to our sensorial systems and notice that even on the larger scale of operation not much is surprisingly new. Our sensory receptors, all of which are composed of thousands if not hundreds of thousands of these nerve cells, work through the same method. They become excited and then pulsate together to be recognized and acted upon by the brain. These receptors are classified according to the type of energy they can detect. Our five senses react to a form of stimuli through energy that is then transmitted into action potential and sent to the central nervous system.[31]

To explain the different types of classifications, I would like to introduce ADM7 the robot. ADM7 needs to see somehow; otherwise he will not be able to walk his way through obstacles, find objects he is supposed to use, and locate his destination when assigned with one. Let's add a video camera to ADM7. The video camera can detect a visible light spectrum and pulsate it to ADM7's CPU. After we spend a great deal of time categorizing obstacles and objects, we are then faced with another problem: we want ADM7 to respond to voice commands. We introduce a microphone to ADM7's circuitry and spend time teaching different frequencies and their combinations in patterns to categorize them. To walk on uneven surfaces, we equip ADM7 with a gyroscope. To avoid danger to his circuitry and harmful gases, we give ADM7 a chemical-smelling machine. We also give him a thermosensor to help him avoid high heat that can melt the lead that holds his chips in place.

All these components, such as the microphone and photosensitive sensors, have been in development years before we were set to build ADM7. They have been used for many other applications, but it so happens that we need them all to work together for our purpose.

We animals are faced with the same problems day-to-day. We are dependent on our vision, smell, taste and hearing for our everyday life. So what are the sensory receptors we use? We share much with ADM7:

- Mechanoreceptors: hearing and balance, stretching
- Photoreceptors: light
- Chemoreceptors: smell and taste
- Thermoreceptors: changes in temperature

These sensors will be in the front lines of this book in later chapters. Although it is not necessary, I would recommend further readings on how they work.

The nervous system, or more so the peripheral nervous system, connects the body to the spinal cord, and the brain also function as a pathway to carry signals from the brain and spinal cord to muscles and glands. This is important, for it is how we function in accordance to our surroundings and send embedded instructions to operate our organs in our body.[32]

The Brain and the Spinal Cord

Coming a long way from its early Aristotle days as a cooling apparatus to the heart,[33] we now know the brain is responsible for our complex thought. Since the beginning of recorded history, the mind has fascinated humans. We were obviously the chosen ones to rule the world, but what made us so different from the rest of the animal kingdom?

The study of the human brain was and still is considered the holy grail of medicine. During the course of history, many hypotheses were raised to explain how the brain works, and several models were introduced as early as the time of the Egyptian empire. Two Egyptian anatomists known as Herophilus and Erasistratus traced mysterious vessels that all led to the brain. Centuries later, an early Greek medical writer and philosopher named Alcmaeon of Croton discovered that there are physical connections leading from the eyes to the brain. This discovery led him to reason that thought is accommodated in the brain.

When seventeenth- and early eighteenth-century anatomists examined the brain, little was obvious. The brain was thought to function as a homogeneously large gland that resembled an inverted tree. This notion came from clever yet morbid experiments, such as that of Jean-Pierre-Marie Florence, which were performed to understand whether different parts of the brain have specific functions to contribute to the whole. In the experiments, Florence systematically removed different parts of the brain from lab animals, and he later examined what remained in function. Florence noted that instead of impairing only selective functions, results showed that all function were at work, only progressively weaker.[34]

In contrast, a Venetian doctor named Franz Gall introduced another model. Gall believed that the surface of the skull denoted the functions in the compartments of the brain. In what was later coined as phrenology, the relationships of sizes of different bumps on the skull were corresponded to different traits of human character. The theory of phrenology was widely accepted in Gall's time but was shortly abandoned when discoveries that contradicted this theory were made.[35]

In later centuries, more elaborate models were introduced. In the 1950s, the brain was thought to be constructed of different evolutionary origins. The triune brain theory, hypothesized by Paul MacLean, introduced the brain as a three-level hierarchical organ. MacLean believed that the "reptilian brain"—what is now referred to as the brain stem—was an ancient relic from our reptilian origins and was responsible for our instinctive behaviors. Next, according to MacLean, the "old mammalian"—now referred to as the limbic system—was developed as our ancestors evolved and is responsible for our emotional behaviors. At the top of the hierarchy, the "new mammalian brain" processed in the cortex, the outer region of the brain. The new brain was responsible for our ability to rationalize and think.[36] This theory provides an intriguing look of how the brain evolved to what is today. We obviously underwent an evolutionary process in which our adaptation to our surroundings demanded more than the instinctual processing that controlled our bodies. However, this theory, which led to some interesting ideas regarding how the human brain works, was somewhat flawed.

This theory, which compared humans to animals, was presented with some undisputable flaws, specifically when compared to the connection between other species' brains and their behaviore. It is possible to believe that all animals, even when evolved through divergent evolutionary conduits, exhibit similar functions. This is not because animals share a common ancestor but because animals independently evolve similar traits to acclimate to similar ecological environments. (This is known as convergent evolution.)

With today's research tools, scientists have a far more comprehensive understanding of how the brain works than ever before. Through medical imaging we learned that the brain has no obvious localized functions; rather different areas of the brain operate in parallel to achieve different functions.[37] Even though the brain is constructed of anatomically distinct regions, it would be ambiguous to think of one distinct region as having one specific self-governing function. More so, we have observed that the brain is capable of reconstructing itself when certain areas are shut down. With outstanding recovery of function, brain cells can gradually learn to take over the role of damaged cells from a particular section and through neural plasticity. Patients who suffer brain damage can sometimes gain function with no or few apparent symptoms.[38] However, it is important to emphasize that one area of the brain, which is the most recent upgrade in our evolutionary development, stands apart—the outer layer of the brain, the cerebral cortex. The cortex is believed to have evolved as we adapted to our surroundings.

The biggest change took place after the environmental transformation that reshaped Africa, then a thriving forest, to a low grassland. The cortex surface area has increased in size while conforming to the skull and folding itself while doing so.[39]

In that sense, the different models introduced in earlier decades all somewhat reconcile with today's understanding of the brain. Although the brain is neither constructed by a three-part reptilian, old mammalian, and a new mammalian brain, nor is it a hierarchically calculating machine with main CPU that controls all, we can see how it has progressed through ancestral evolution.

Because this book deals with what we define as "thought" and what reasoning is, a much deeper look at the cortex must take place.

The Cerebral Cortex

Throughout many years of research and observation, we concluded that the mind is located in our body, to be more accurate in our brain. We make the decisions and actions that define our individuality or character while we are awake and cognitively conscious to make them. Hence, something should be responsible for our actions. When one questions the existence of the human soul, a controversial debate often follows. Is a soul so different from anything else we have observed through the years of medical research?

If we came to the understanding that the earth is not flat and that the universe does not revolve around us, should we not also detach ourselves from the notion that the soul is a mysterious creature that dwells in our body? When a phenomenon cannot be explained, myths and legends are born, and, however fascinating they might be to write or make movies about, they should not be the cornerstone to our knowledge. Even though there are many holes in human knowledge, if we are able to modify something with the knowledge of the result of the change, we have more understanding of it then we ever had before. Electricity is a good example. Even though we do not fully understand energy, we can harness it to meet our needs. A chronically depressed human can now take a chemical that will make him or her less depressed because we now understand that this person might be deprived from this naturally produced chemical. Therefore, if a human's personality can be modified through medication, it is probably time for us to let go of the age old belief that a soul may exists. By this means we are to deduce that we now have a better understanding than ever before to how humans have come to be what they are.

Where does this thought and state of self all take place? The answer is found in the cerebral cortex. To the naked eye, the cerebral cortex is a stretchy mesh of matter about two millimeters thick. Often referred to as gray matter because of its color the cerebral cortex is formed by a complex network of neurons and axons that connect different regions of the central nervous system. Unlike the central nervous system, a myelin sheath does not envelope the neuron fibers.

The surface of the cortex is folded and can be divided into separate areas and layers. The two main groups of layers, the hippocampus and the neo-cortex, which is more recent in evolution, are composed of five layers and six layers, respectively. The cerebral cortex receives messages from stimuli through sensorial inputs, yet its main source of stimulation is itself. Areas of the cortex that receive particular information can be categorized according to different primary sensory areas, which receive inputs from the thalamus. Visual cortex, auditory cortex, and somatosensory cortex are all stimulated when they receive impulses from vision, audition, and touch receptors.[40]

To react upon stimuli, we operate our body through voluntary movements by activating motor areas. A contraction of a muscle occurs when a signal is sent from the brain along the spinal cord. Before the signal is sent to the spinal cord and to the muscle, the message goes through the supplementary motor areas: the premotor cortex, which selects between voluntary movements, and the primary motor cortex, which executes them. These operations have further classifications that separate higher-order instructions of self-generated movements from others that dictate how to move in the surrounding space.[41] Memory takes part in instigating and processing these motor functions.

When information is processed and when brain activity takes place, these areas work together in the process we might call thought. The magic takes place when messages integrate with additional information in the three major associative areas that make up much of the cortex. Areas of the cortex are comprised of three major association groups:

1. Partial, temporal, and occipital lobes are all located in the posterior part of the brain and are involved with our perception of our relative environment. What we see, hear, and touch is processed in these areas to give us sense of position and proportion to our surroundings.

2. Frontal lobes, also referred to as the prefrontal association complex, are the largest part of each cerebral hemisphere. The frontal lobes play a key role in many of the functions that define us as thinkers. Of course, there are distinguished asymmetrical differences in the frontal lobes; one is

involved in controlling language-related problem solving, and the other in nonverbal problem solving. However both are involved in processing and association.[42] While we have gained much of the knowledge by observing mentally healthy and unhealthy humans and animals, we can observe that damage to the frontal lobe can result in a wide variety of symptoms. MRI imaging in case studies have shown that following an injury or birth defect, functions—such as memory, language, judgment, control over impulses, and social behaviors—can undergo a severe change from what we define as normal. These symptoms can include social behavior changes, such as interacting with others in a social setting, to the inability to recognize shapes or even notice moving objects.[43] A question arises—how can one area of the brain be responsible for such a variety of symptoms? To answer that is truly to answer how the brain works, but as we shall see in later chapters the key word will be *pattern recognition memory.*

When an area is responsible for receiving information, a pattern of signals, from other sensory parts and is accountable for cataloging this information for further processing, damage will have a tremendous impact on various functions. To illustrate how damage to a small section of an associative area can cause problems, let us look again to the technology world.

Let's assume you have a computer with which you surf the Internet. Through unfortunate circumstances, a malicious virus has taken residence in your computer and restricted your computer to play an MP3 file. The virus was launched to your computer's operating system and deleted one tiny DLL (dynamic-link library) file. This file was, in a broad sense, responsible for knowing how to work with MP3 format. Before the operating system can play a sound, it has to first know that a file should be dealt with through a program that understands sound formats, such as the media player. This requires at least one type of association. Then the computer has to remember how to interpret the ones and zeros that compile this compressed file format to a symphony of information that will eventually pass through to the sound card, which will eventually guide how fast to push the speaker's membranes and create sound. After contracting the virus, when you click on an MP3 file, you will get an error, which will inform you that the system cannot read this file. you will be able, however, to play CDs and other sound formats.

In contrast, a much more destructive virus can strike your computer. This one would take down the computer's ability to associate between DLLs. This time not only will you be unable to play MP3 files, but you will have severe problems with all file formats, such as word processing, images, and many others file formats vital to the operation of the computer. In that sense, your computer will be virtually useless until you install a new operating system.

I have illustrated two extremes, and damage to the operating system can come in many different shades on the spectrum. As associating machines, humans work just like anything else that require associations, and, when damage to the process that deals with association takes place, it will interfere with the chain that is responsible for problem solving.

3. The limbic associative areas (or the limbic system) are a group of brain structures that are common to all mammals and are critical for normal human functioning. These structures include the hippocampus, amygdala, and olfactory cortex, along with various connection structures. If we were to associate an area of the brain that is responsible for emotion and memory, it would be the limbic system. Although we are currently unable to pinpoint where memory is stored (it is most likely ingrained throughout many brain areas), we know where it is formed and catalogued into a coherent existence. That is, when memory is called upon in the brain, an area, such as the hippocampus, cross-references incredible amounts of patterns that have been stored and creates an image that is our tool for recognizing familiar objects or surroundings and storing new patterns for future cross-referencing.[44] Another striking feature of a limbic member, the olfactory system is unlike most of the structures of the sensory system in that it does not have to pass through the thalamus before reaching the cortex.[45] This is where the three-level hierarchical brain theory surfaces. The sense of smell has been with us since the dawn of our primeval evolution. Playing a vital role in almost all aspects that were to protect us as living species, the sense of smell was responsible for humans' ability to find or avoid different foods, to locate dangerous enemy species from afar, and to find a mate in the wild, all of which our existence would not have continued without. The olfactory system begins in the roof of the nasal cavity and with its millions of receptors is capable of detecting thousands of differ-

ent odors. This four-layered area is a primitive old structure compared to the six-layered cortex structures that developed later.

The amygdala has its own role for our survival. In an attempt to survive an animal must smell danger. To catalogue this smell, an animal must remember that this sense is danger and then activate an adaptive response to it. The amygdala gets sensory input from most of the associative areas. Visual, auditory, and somatosensory areas send highly processed pattern information, which is catalogued in the amygdala and provokes a set of autonomic responses.[46] As well as memory, I will dedicate an entire section to emotion and its possible roots in the limbic system. Emotions, such as fear, have a tremendous effect on reasoning and logic.

To conclude this section, imagine a musical piece that is performed by an unconventional orchestra. The piece has called upon all the professional performers in the world (who have agreed to cooperate together) to assemble in a specially built stadium, in which the seats are filled with more and more performers. The composer divides the newly formed super orchestra into many smaller ones, each constructed by its own sections of wind, string, brass, and percussion sections. The smaller groups all play different familiar and less familiar pieces of many great composers.

Even though the performers are all playing different pieces in different keys, the grand piece is arranged in a way that, when a listener hears it being played, harmonious coherent music will be audible. All this happens instantaneously and spontaneously. For example, a listener asks a conductor to play his favorite tune, and he complies. He pulls the entire notation, and the performers play it. While they play, the group next to them plays a different tune. The well-trained conductor will match the tune's key and tempo to yet another group that is playing a different tune, and so on.

Some of the tunes have been written years before and appear on endless sheets of music that are filled with forms, such as concertos to rondos, that are composed of smaller sections of musical patterns such as 4, 8 and 16 musical sentences. As unbelievable as it may seem for this to actually occur in real life, it happens every moment in conscious thought. In our specially constructed stadium with our own performers playing notations that were composed through millions of years of evolution, we play music that is drawn from many regions of our brain but comes out as a thought or action that helps us interact with our surroundings. In the past chapters, we have considered some of these performers and conductors. Now we must look at our own musical notation in detail.

Chapter IV: Neural Alphabet and the Vocabulary of the Mind

Now that we have taken our first steps into understanding the brain, it is time to start walking towards the understanding of our native language. We have established that our body communicates through electrical signaling; that is, single neurons send messages to each other. I believe that taking the top-to-bottom approach in completing our understanding of the brain is the preeminent method adapted in this chapter. We must first learn how association memory is created and stored in the brain before we can see what pattern association memory is and how it works. This approach is similar to learning that there is a glass sitting on a table, then learning that this glass can hold liquids, then understanding how to put liquids in the glass, how liquids are stored in it, and finally learning about the liquids that we have poured into the glass. A far better example can be drawn from the world of technology. We shall first learn how it is possible to store information on a media (memory media, from DVD to magnetic tape) before we learn how to program the operating system.

Memory

Memory is important. If there is one feature of our brain that we can safely say is responsible for our exceptionality, is our unique plastic memory (plasticity). Although other mammals have memory, our ability to use our memories to adapt to our surroundings makes us unquestionably unique. Combining our capable memory to our unique hands and voice boxes, and we can see how humans can create complex structures and speak intelligently.

Children, at some point in their lives, ask why they have to learn history, and we always tell them, "Those who don't know history are doomed to repeat it!" This answer is true. Memory is central to our survival by helping us avoid danger. The knowledge of the parent protects the newborn, and only through remembering and imitating the parent will the young be able to leave their supervised protection and have offspring. Although genetically embedded memory does exist, there is great importance in what we learn from experience and observation.

Our brain in a sense is a calculating system. However, compared to the transistors in a computer chip, neurons are quite slow. A single neuron can collect inputs from its synapses and spike them to other neurons in about five milliseconds. More so, the neuron can perform that and reset itself two hundred times per second. Although some might consider this speed fast, a personal computer can process information a million times faster than that.[47] The measurement of computer performance is measured in FLOPS (an acronym for FLoating point Operations Per Second). Nowadays, teraFLOPS (which are one trillion operations per second) can be seen in the consumer video game market, and professional computers can outperform this speed with ease.

How does the brain give us the impression of performing with greater speed? An argument can be made that there are billions of these relatively slow functioning neurons all work in parallel. The brain is then capable of processing at much faster speeds than a supercomputer because there are billions of small processors all working at the same time.[48] However, although the brain processes information in its associative areas in parallel, it does not do so as a net of billions of mini neural processors running in parallel to calculate a large problem. The human brain, for example, can recognize an object in less than a second. If you were asked to press a button every time you see a horse in a group of pictures that are presented to you, it would take you less than a second to recognize the horse. A computer, on the other hand, will have much more difficulty in processing that and will probably fail when presented with pictures of similar animals, such as a zebra or even a Great Dane, for that matter. The number of steps to achieve this recognition in the brain is limited to the confinement of the speed of the neuron. However the brain solves the problem much faster than a computer. If so, there must be fewer steps of calculation in this recognition process than that of a computer.[49]

Before we propose the secret of brainpower, I will provide some more examples of daily activities that might seem as though they require calculation but do not require an advanced supercomputer to occur. Walking, for example, is an achievement engineers have been trying to perfect in robotic legs. The task of mapping an area and adjusting the mechanics of robotic legs to walk from point A to B requires a great deal processing, and few companies have achieved even a maladroit bipedal model.

Are we supercomputers who can process billions of variables? The answer to this is simply no. Computers process much faster than the human brain. Quantum computers will most likely become so fast that the machine will surpass us when performing any calculation. However, even with all this computing speed, computers will still have many problems to solve before being able to walk or run as we do, with little thought and without scanning every variable. So how do we do it?

We simply remember how. From learning our first baby steps, we progress through many years of learning how to walk and run with less clumsiness. We remember how to take different steps on different occasions. Even as adults we will have much less trouble running down or up stairs we have traveled hundreds of times than steps we are taking for the first time. With little effort, we remember where to place our feet because we recognize the distance from our feet to the stair. If you ever missed a step, a strange feeling probably occurred right before the expected feeling of pain when hitting the floor. Something had thrown the brain so far from what it expected that a sense of high alert had been flagged and action to protect the body had to take place—unfortunately, not always with sufficient reaction time.

For those of us who enjoy a sport such as tennis or squash can surely testify that only through time and practice can someone succeed in a tournament. Although hand-eye coordination is crucial, we also have to learn how to react to a ball moving from left or right. The most wonderful feature of our brain is that instead of memorizing thousands of moves specifically for this game or for this certain opponent, our brain plasticity takes over by adjusting our movements. Processing is required, however, not as much calculating as a robot that must calculate the exact ballistics of every ball in question.

There could be endless examples of daily activities that require the recall of memory. However, I will only entertain one more. The notion of conveying examples is by itself a good example. The best way to learn new information is either to experience it or to learn it from examples we can relate to. Through examples we can assimilate new information to our memory by associating it with what we already know.

The question remains, how does a chemical-based electro signaling system store information? To solve this, scientists have first looked at simplified organisms. Throughout its existence, life as we perceive it has used the same methods to solve problems. Living organisms have used all the resources that the universe provided us with to deal with our own survival. We have not actually studied how chemical compounds store information then created our own sophisticated structure, but, through millions of years of evolution, these methods came about naturally.

The work of Nobel prizewinner Eric Kandel has opened a path to much exploration and understanding of how the brain stores information through changes at the synapses in a neural circuit. Kandel's research focused on a simple organism, a sea creature called *Aplysia californica*—the California sea slug. This animal, whose diet is comprised mostly of seaweed, has a simple nervous system compared to ours, which gives scientists an advantage in studying the function of its nerve cells. Although small in number, the *Aplysia's* nerve cells are quite large and distinctive. This helps scientists to compare trained and untrained slugs at their neuronal and molecular levels and, by doing so, to observe what changes their training instigated.[50]

With the daunting task of mapping the nervous system of the *Aplysia* cell by cell, training took place, targeting the gill and siphon withdrawal reflex, a hard-wired defensive reflex of the sea slug that causes its siphon and gill to retract when the animal is disturbed to protect the flimsy respiratory organ from potential damage. The experiments showed that after about ten times of lightly touching the siphon with a little brush, the slug was conditioned to ignore the now-familiar stimulus, and the creature showed little to no withdrawal of the siphon and gill.

This experiment, along with others, has shown that this simple organism can undergo various learning procedures, including habituation, sensitization, and conditioning, which are common to higher-level organisms, despite the simplicity of its neural circuitry. The duration of the memory to the habituation depends on the extent of the training—a first clue to differentiate short- to long-term memory.

Now that the scientists had established that this simple animal could create and hold memory, they needed to understand what changes, if any, were noticeable in its nerve cells when memories were formed. In other words, something must have been modified to contribute to the formation of memory, especially when it affected a neural circuit that was impeccably predetermined and prewired. When the slug was dissected and the sensory and motor cells were isolated, microelectrodes picked up that the strength of the synaptic connection in the sensory neurons dramatically weakened when electrically stimulated with the same pulsation of an intact animal. Kandel and his colleagues found that this weakening was a result of a decrease in the number of packets of neurotransmitters that were regulated through action potential of the cell. The experiment showed that not only short-term but also long-term memory could be formed in this way. Longer and more extensive training caused much greater changes in the synapse and in some cases led to the disruption of most of the previous connections. Another experiment performed by the team showed that in complete contrast, when training the slug to become more sensitive through many rounds of electrical shocks to its tail, the sensitization for the gill and siphon withdrawal reflex lead to an increase in the number of synaptic connections. This entire process, whereby strengths of synaptic connections are altered, clearly supports that synaptic connections are not fixed but can be adapted by any form of training. This adaptation of synaptic connection is called synaptic plasticity. Plasticity is a process in which the brain adapts to different scenarios using learned patterns. Again, we shall see how a pattern can be used to solve a variety of completely isolated problems that are by themselves unique features of what we perceive as existence, but are in the grand sense interchangeable.

In a highly adaptable nervous system, some mechanism has to provide the foundation to this plasticity. Although this biological process has not been fully determined, one apparatus, long-term potentiation (LTP) and its opposing process, long-term depression, is believed to contribute most. First observed in the hippocampus of a laboratory rabbit by Terje Lomo and Tim Bliss in 1966 in Oslo, Norway, the process was discovered unexpectedly after stimulating the perforant pathways with a brief set of electrical impulses. As expected, a single pulse delivered to the perforant pathway resulted in an excitatory postsynaptic potential (EPSP) in the dentate gyrus. What they did not expect was that a set of high-frequency stimuli resulted in a stronger, more prolonged EPSP that could last as long as 120 days in an animal. Since the publishing of their research, LTP became popular in the field of memory research. LTP has been observed, in different forms, in many other brain structures, including the amygdala and the cerebral cortex, to name a few. LTP is dependent on messenger ribonucleic acid (mRNA). A template for protein synthesis, mRNA is manufactured from a DNA template during transcription, in which generic information mediates from the cell nucleus to ribosome in the cytoplasm. After a signal is sent from a synapse to a nucleus, proteins that can alter these protein networks are produced with certain gene expressions that regulate changes of synaptic strength.[51]

Formation of Longer-Enduring Memory

Although we cannot and should not try to isolate one area of the brain and flag it as the center that holds vast and long-lasting memory, we do know what part of the brain fabricates this memory, and we are beginning to understand exactly how it is manufactured.

In 1953, Dr. William Scoville, a neurosurgeon at the Hartford Hospital in Connecticut, performed a historic operation. The patient, a twenty-seven-year-old known as H. M. had been experiencing severe and frequent epileptic seizures. In the procedure, Scoville removed a complex of brain structures that included approximately two-thirds of the hippocampus, the amygdala, and neighboring areas of temporal cortex. Immediately following the procedure, it was clear that something had gone wrong. Although H. M.'s seizures were somewhat subsided, he now suffered from severe anterograde amnesia. His short-term memory was intact, but he could not consign new events to long-term memory. The procedure left him trapped in time. He could not remember any events that happened minutes earlier or where he was. In short, he was literally living in the moment.

It is worth mentioning that even though H. M. suffered from moderate retrograde amnesia, he could remember much of his life before the surgery. H. M. could remember some of his childhood and teenage events in great detail. His ability to learn new procedures was still undamaged. However, he would not remember learning them. That is, his long-term procedural memory could still form in his brain and he was able to learn new motor skills, even though he would not remember the process of learning them.[52][53] To illustrate his condition, imagine playing a video game. This video game requires high levels of hand-eye coordination. Each time you play the video game, you train your response time to an event that happens on the screen, such as a bad guy appearing with a weapon, and your react to that event, pressing a button to defend from his attack and attacking him with another button. After many attempts at playing the game, you will eventually get better. After playing this video game repeatedly, you can look back and laugh at how awful you were when you started. If you were a patient like H. M., you would continue to get better; however, each time you play the video game it would be unfamiliar, as if it was your first time playing it. Moreover, after a minute or two, you would not remember why you were holding a controller, sitting in front of a video screen, or how you even got there in the first place.

Throughout years of scientific testing, many other experiments were held to explore the secrets of the brain. After H. M.'s famous case, Canadian neurosurgeon Wilder Penfield studied in the 1940s and 1950s the brains of epileptic patients. Wilder's main goal was to pinpoint the areas responsible for the patients' seizures. Using electrodes in direct contact with the patient's brain, he triggered different regions while the patient remained conscious. The procedure was not painful to the patient because the brain does not have pain receptors. The patient could stay awake and report to the surgeon during the entire procedure.

The procedure was rudimentary yet highly original for its time. While stimulating the brain with electrical probes, Wilder could observe what regions he was probing through the patients' reactions. He knew, for example, that he was stimulating the auditory cortex when the patient reported hearing voices or music, or that he was stimulating a motor zone when the patient exhibited a hand or leg twitch. While probing the temporal lobes, patients reported vivid dreamy states of events that occurred in earlier experiences. The patients almost experienced these events in flashbacks, similar to a movie.[54]

Although these experiments have contributed to our understanding of where memory might be located, by no means have they shown us that the temporal lobes are where we store all of our memories. The notion that we distribute our memory throughout the brain still stands unchallenged.

Where H. M. and patients of his condition have given us insight to how a brain's particular areas may be linked to the process of storing new information, there has been ongoing research to locate the manner in which long-term memory is created. Recent research, which was published in 2006, found sufficient evidence that a specific gene plays a key role in the consolidation of memory storage. Long-term potentiation (LTP) was mentioned earlier. It was also mentioned that genes are released to regulate the strength of the synaptic connections. Activity regulated genes Arc (Arc/Arg3.1 activity-regulated cytoskeleton-associated protein) are among them. The group of researchers created knockout mice, which carried a void mutation of Arc and were unable to self produce the correct form of this gene. The results were similar to that of H. M. Although their short-term memory was unaffected, the mice's long-term memory was deeply impaired. The mice were subjected to several tests that examined long-term fear-related memory, long-term object recognition, conditioned taste aversion memory, and spatial learning strategies. The results of the tests all yielded a unanimous result. Although the early phase of LTP was greatly enhanced, the late phase failed to consolidate, and, after approximately sixty minutes, no potentiation remained.[55]

Although we have barely tapped into a vast and growing field of knowledge in the research of memory, we have established that memories are not miraculous, per se. Moreover, we have observed that all organisms are not a product of intelligent design but are a result of evolutionary adaptations that are made within the context of a building block in nature. In later chapters, we shall see how this notion will reappear again on a possibly much larger scale.

The Neural Code

Because we have established that we store memory in its verious types and that our psychological processes are physical, we can go on to what we do with the information that is absorbed by our various inputs. This section will touch upon the subject of cognitive perception association but will also present a skeptic philosophical predicament that might question our entire knowledge of the matter.

From the moment we receive information to the moment we react to it, a certain amount of processing is required. To understand our surroundings we must first detect them in some way then, with the help of several apparatus, process the information and react. The amount of information that passes through a brain at any given moment is so immense that most computers would be unable to process it. How then does the brain naturally deal with such a vast amount of information? How can one, for example drive a car while preparing himself for a meeting in the afternoon, and at the same time be bothered by some event that occurred last week, while listening to music on the radio? It was mentioned that memory is the basic method for an animal to learn and react to its environment, now is the right time to discuss what it is that we actually write into these neural networks that we call a brain.

Normality as Represented by Patterns

One of the best tools humans have to cut processing time is to organize objects into categories. By doing so, we can access these objects at a much faster speed then if we deposit them in one massive pile. Objects are placed with other objects if they share some trait that groups them together. Consider a toy box that has different holes in it and separate parts that matches these holes. This hollow box has a circle, a square, a triangle, and a star cut into it, and the purpose of the game is for the child to take a circular piece, for example, locate its corresponding circle hole, and insert it to the hollowed box. Behold the first steps into voluntary cognitive categorization. Other educational toy can extend not only to shapes but colors (placing all the red balls in a red basket), animals (placing all the lizards in the lizard cave), and so on. When the child is asked to retrieve a red ball, he or she will reach for the red basket and will promptly retrieve the object.

With this simple task numerous set of voluntary and non-voluntary cognitive association processes begin. The child had to understand language to understand a name of a shape or a color name (which is merely an assigned name for different light spectrums our eyes detect). Each process is a complex set of associations that requires pattern recognition and association.

A pattern is anything that can and will perceptually occur at least twice. Arguably, anything that happens once can still be categorized in its own class, yet the whole purpose of placing this unique trait that occurs only once into its unique category is for identifying future similar traits. Mind you, nature is far less simple than that. A ball in a color we have never seen or heard of before is still a ball and can be placed into the family of balls. Not only its shape, but its material, weight, and many other traits can be associated with other categories in which it can be placed.

Our brains process everything through these numerous patterns. When a test subject while looking at picture of a spoon is studied under an MRI, it will be obvious that part of the apparent visual stimuli and other areas of the brain excite at the same time. If one will examine further, one might find that memory will come to play in every aspect of the process after the visual stimulation. The spoon shape and material will have to be identified in the subject's brain. On the macro scale when the subject is looking at the spoon, he might be reminded of what cutlery is or what it is made of (is it metallic or is it plastic?), or he may even entertain the thought that there is no spoon, as he saw in a movie once. This memory recall might lead to an entire set of other associative memories that involve dinner with the family or a special night with a friend. On the other hand, it is important to understand that for each and every bit of this memory, micropatterns are used to process the recall of this associative memory.

To understand the full complexity of this process, we need to dig deeper into the circumstances in which a pattern might be used in both a recognition process and in a stimulus-response rule (response arc). A bottom-up approach will begin by attacking a first contact of a receptor cell, then tracking it as it sends information for further processing. Because we began with visual stimuli, we can start our excavation by tracking visual pathways and their communication to our brain.

Our eye, a complex organ composed of many working parts, is our contact to visual stimuli from our surroundings that are passed on to the occipital lobe. In its core, two types of specialized nerve endings convert light into electrochemical signals (photo transduction). These photoreceptor cells, rods and cones, respond differently to the electromagnetic spectrum. Rods are used for low light vision, whereas three different types of cones are responsive to daylight color vision. The ratio of cones to rods photoreceptors varies in different animals, mostly depending on their nocturnal or daily habits.[56]

Before we consider recognition of shapes, our first contact will be that of color. Let's take a glimpse at how color is understood through the process of photo transduction in cone cells. First, consider what exactly color is. Color is the frequencies of visible light. The photoreceptors of the human eye can detect electromagnetic radiation in the range of about 750 nm (seen as red) to about 380 nm (seen as violet). Higher frequencies of light provide us X-rays, gamma rays, and cosmic radiation. In the lower spectrum, infrared, microwaves, and radio waves have all been harnessed for communication.[57] Some insects, such as bees, are able to perceive infrared light, another fine example of evolution, because it gives them a direct advantage in finding the nectar in flowers.

Long before the invention of the lightbulb or even our exploitation of fire, animals were dependent on the sun, which produces most of the radiation of the electromagnetic spectrum. Our ability to see color begins the moment the sun's radiation bounces off objects that surround us. The properties of their surfaces affect the color of the objects. The property of the surface of a red ball, for example, projects outward light waves, which are around 750 nm. In contrast, a black object will absorb most of the light it receives. (This is why you should avoid wearing dark clothes on a hot summer day.)

L-cones (long red wavelengths), M-cones (medium green wavelengths), and S-cones (short blue wavelengths) collect the light that bounces off an object's surface. The different cone types have a slightly different protein, which is called opsin. In the process of photo transduction, from these three types of cells the output and intensity will be coded and later pinpointed as a color.[58] Painters were probably the first to understand the results of mixing red, green, and blue colors. With these three colors, one can create a wide selection of colors. The amount of green-to-blue or blue-to-red ratio in a mixture, for example, can produce a variety of colors. Our electronics make full use of the same methods. Digital cameras and photo-editing software uses the same RGB methods that have been in existence in nature for as long as animals have been living on this planet.

After the photoreceptors absorb light photons, data from the membrane's potential of each individualy excited cell is collected and with further interpretation can form a pattern that the brain can identify.[59] I know a ball is red after my brain collects information from my visual cortex, which brings the electrical signal and codes it to a pattern. I will put this information together with my English lessons in school, and I can access another vast collection of memory of phonetics. After integrating the patterns, I have the almost instantaneous thought: "This object is red."

Now that we have a fair understanding of color, we can go further in our excavation of a visual contact. While we categorize an object's color, we strive to find out what we are ultimately looking at. Not only have we unconsciously used patterns to categorize our surroundings, our mind also makes every effort to use them consciously. Let's extend our ball example further. Instead of a regular spherical ball, we are now looking at a red American football. To make things more complicated, let us consider that our test subject has yet to see an American football. However, the photo we show to our test subject captures the moment a man dressed in sports clothes (a number is apparent, together with vibrant colors and markings on the shirt) with protective gear is nearly catching that strange object as it flies through the air. This will make it obvious to our test subject that he is looking at a recreation ball, even though it may not have the standard shape of a ball. There was much reasoning behind the conclusion. Sports gear and the athlete's expression and posture provided enough evidence that the oddly shaped artifact is in fact a recreation ball. Should our test subject be more generally learned, he might have also noticed the aerodynamic shape of the object.

Visual, touch, and auditory perception in the brain make use of change through time. For example, let's say you were asked to insert your hand into a hole in a box and place your finger on the surface of an object that is located inside the box without moving your finger. Could you identify the surface you are touching? We can start identifying the object only when we move our finger and notice the changes of the corners and the surface materials through time. This gives us an ability to guess size and shape and to reference the inputs to our memory. Our brain can identify an object by looking at it almost instantaneously, yet it requires the eyes to move rapidly and to look at different sections to form a whole.

Coming back to the football, we can concur that the ball is indeed oddly shaped. However, after the eye examined the many areas of the picture that were almost instantaneously associated with the memory of shapes and colors of sportswear and protective gear, our test subject will easily conclude that this is an actual ball. Therefore, this change in input over time works both for the macro and the micro scales of events.

Please consider the full process. We could have simply placed a picture with a white backdrop of the ball, and the test subject may or may not have identified it by its three-dimensional shape, and its leather material. For that matter, hand our test subject the oddly shaped ball, and he have quickly conclude it is a ball after touching it and finding it is filled with air, another process that requires a tremendous amount of pattern recognition.

The process in which our test subject identified the ball is the practice of gathering information (an athlete is in the picture and perhaps a green field), recalling memory (a number on a shirt and a logo can indicate a sports shirt, and sports are often played on grassy fields), and ultimately an answer.

Our brain operates through pattern recognition. It is not a coincidence that our thought imitates how the brain work on a microscale. To understand the world around us, we have built a set of patterns and grouped them together to form a picture of what it is we see, hear, touch, taste or smell. We use our brain to evaluate our surroundings. By deriving a reliable generalization from observation and recall of memory, we can predict what might happen next, and more importantly how we should act next.

This particularly key process of inductive reasoning is formally known as logic. This tool of our brain could possibly be the most important tool for any animal's continued existence. Where did it all start? How far did it evolve? Many more questions will arise from this subject. We will try to discuss them in the following chapters.

UNIT III

Chapter V: The Origins of Reason

We are all dedicated to finding patterns. Observing patterns will directly help us to anticipate what can or will happen next or in the future. When we know what may happen next, we can kick-start our defense mechanisms much earlier than if we encounter an anomaly for the first time. Reason, then, is essential to survival.

If we recall, the *Aplysia* was tested to determine how memory is created. The animal was conditioned after undergoing several tests. A simple yet immensely important process occurred. The animal retracted its gills once it was stimulated. This was an obvious effect. It had to be more protective of its vital organ once danger was eminent. When this stimulation occurred many more times with no harm to its body, the slug stopped showing a sensitive reaction to the same stimulation.[60] After all, there was no need to contract the gill and waste energy if there was no harm. In short, the animal had memorized that the stimulation was not negative. We must face a question that arises from this example: why did the sea slug want to save itself? The answer is simple. However, it is the gateway to many more questions.

As far as we know, the sea slug did not choose to save itself. The sea slug was simply acting through an automatic response, a reflex, a hardwired neural action to protect itself, a mechanism that is solely there for its survival, an instinct.

Reflexes and Instincts

Reflexes are automatic neuromuscular reactions to specific stimuli. Both simple and unlearned reflexes can be observed in all animals. In virtually all cases, neurons govern reflexes. (There is one known organism, ciliated protozoa, that is not governed by this rule. The single-celled organism has no neurons; however, it does exhibit reflexive behaviors.) The more complex the reflex behavior is, the more neurons are required to achieve it, which means more synaptic connections will be involved (the reflex will be mediated through a neural pathway called a reflex arc). A large group of neurons will be connected by specific synapses, form-

45

ing functional units to the nervous system. In most cases, reflex actions occur very quickly by activating motor neurons in the spinal cord without first being routed through the brain (although it will receive an input at roughly the same time as the reflex takes place).[61] We mentioned earlier that these biological circuits require a sensory neuron at one end and effector neurons at another. These circuits can operate regardless of the animal's main processing unit (what we call the brain). However, they are nonetheless an important and integrated part of it. For example, most muscles can be activated through a reflex or through a signal sent from the motor area of the brain. Muscles are operating units in their own right but do take instructions when they are called upon to.

For the purpose of further illustration, let's construct a reflex robot. We will need simple parts that consist of a light sensor, a controller board for the electronics with all the necessary chips and their connectors, an assortment of electrical wires, a small motor, some acrylic glass, and a battery.

With the help of careful cutting and gluing, we will construct a T-shape stand and an I-shape arm. We will attach the control board to the base of the stand, and to the arm we will connect the electrical motor, which also be attached to the top of the stand on one end of the arm and the light sensor will be attached to the other end of the arm. After connecting the necessary wires and the battery, it is time to give our creation some instructions.

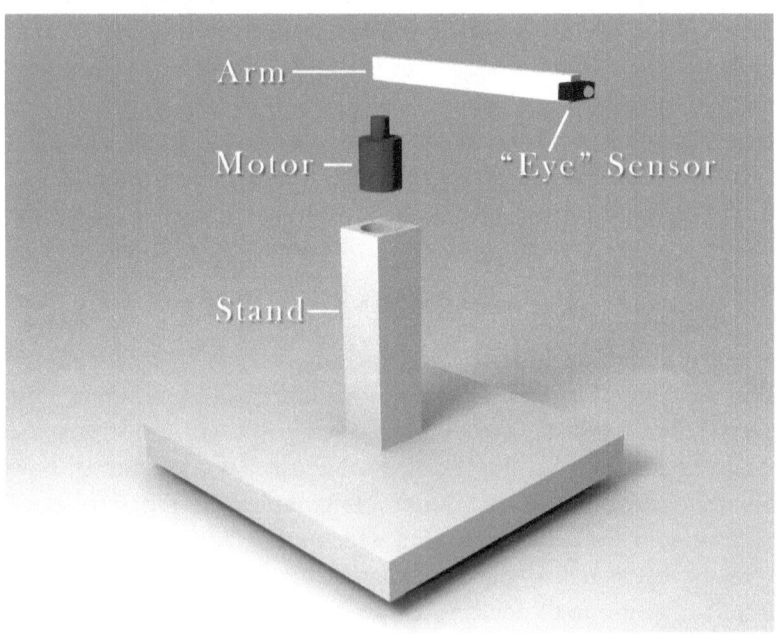

The goal for this example will be to destroy its moving parts. However, it is now time to help our little robot with some programming. We will instruct the control board to ignite the motor whenever light changes around the sensor. This will cause the arm to move in a circular motion when our robot sees a change. After a quick circular movement, the arm will stop at a time we programmed it, preferably less than a full circulation.

Now it is time for our experiment to begin. We turn the robot on, and it appears to be motionless. We now swing our arm toward the robotic arm. Suddenly, the little robot moves its arm to a different spot. We have created a reflex robot. The robot notices light change and mindlessly reacts according to what it has been instructed to do. When "danger" is near, the robot's arm escapes.

The robot did not know that it is facing danger. However, we achieved our goal. If it did not have that mechanism, the robot would have suffered from potential danger and this is where millions of years of evolution play an integral role in the evolution of reflexes to develop brilliant survival machenisms.

It is a simple concept if you look at it from a Darwinian perspective. Since the dawn of biological creatures, organisms that have had sufficient defense or attack mechanisms have survived, dominating others who had less sufficient survival apparatuses. A survival instinct is indeed a hardwired pattern of behavior. It is there not by chance, because only animals that developed it actually survived. It could have originated as a mutation of a microbiological organism. However, it proved to be so effective that the survival instinct continued to exist.

Ever since Darwin suggested our origins to the world, evolution theory has been in the spotlight of the scientific world. Before we continue with survival instinct, let's look at how evolution actually occurs.

Our best weapon against antievolutionists, a part of sound reasoning and proper education, is sheer evidence. Because it can be hard to convince those who do not believe in evolution that we have evolved from apes and lesser species—mostly because of the time it takes to actually see it happen in nature—we can focus on a different organism that can mutate and evolve in a matter of days—the human immunodeficiency virus, also known as HIV.

HIV replicates quickly. While replicating itself so many times, mistakes in the genetic code happen to a few of the millions of replicated viruses. One of the difficulties with AIDS treatment is the rapid mutation of AIDS, making it impossible for one drug to be effective on the virus (a process also known as antigenic drift). A drug can be effective on one form of the virus, but another form will be immune to the treatment. This mutated form of the virus will replicate itself many times over, and the patient will have to be tested again and refitted with a new drug to address the mutation, which will be an ongoing battle.[62]

This is not far from how Darwin's theory of the survival of the fittest works. Animals that were more capable of adapting to their environment survived. We can be certain that all animals that exist today have some form of a hardwired instinct that keeps them surviving as a species. Survival instinct is behind why we are hungry, cold, and scared as infants, and why we feel the need to procreate later as we develop. Is it possible that the survival instinct is behind much more complex behaviors, some of which we have dedicated hundreds of years of study in philosophy and psychology? Is it possible that, reasoning, and ultimately human thought instigated from a hardwired instinct? As we shall observe, it must be the case.

What Is Reason?

We hear about reason, we discuss it, and we certainly use it every day of our life. I cannot recall how many times I have asked myself why—usually in reference to the reason behind what I am currently dealing with. From a young age we question why" more prominently than any other question. We are constantly obsessed with why things happen, because we are unconsciously infatuated with how we should adapt to our surroundings.

Early philosopherse dedicated their lives to finding what it is that makes humans different from other animals. Although much of their research is now understood to be a noble, artistic attempt, the most important result that came from their efforts is the discovery—or for the lack of a better term, acceptance—of the reasoning process.

One of the forefathers of European philosophy, René Descartes, states the following, "*Dubito, ergo cogito, ergo, sum,*" which is Latin for "I doubt, therefore I think, therefore, I am." Regardless that numerous books and papers have been written with criticism for this statement and that events later led the philosopher to change the statement himself, there is a profound insight in this statement—the acknowledgment of thought. Yet this did not provide any evidence for the superiority of humankind. The early philosophers would not solve this question of superiority, but only through the fruition of modern science can we finally answer this question.

In the dictionary, reason has various meanings, all pointing into the same direction. Some involve the theory of justification (in a group containing motivation, decision and action, which are all underlining parts of logic, an analytical thought that is the premise of intelligence); others deal with mental state. The *Oxford Dictionary* describes reason as "a faculty of the human mind that enables logical inferences to be made and rational arguments to be undertaken to understand the world and solve problems."[63]

Both classical and modern philosophies have paved the way to the study of epistemology, the theory of knowledge. This field of philosophy is far from being entirely academic and has provided practical applications to many other fields, especially dealing with justification. Some common applications are sciences, including mathematics, medicine, Artificial intelligence, and psychology to name a few; also other fields of law, product testing, and literature. There are enormous accounts in which the theory of justification can be applied. They all have something in common: rules of behaviors—in particular, patterns.[64]

Two distinctive schools of thought about where knowledge comes from divided the early philosophers. The rationalists claimed that all knowledge is a priori, knowledge that exists and has always existed. On the other hand, empiricists argued that knowledge is a posteriori, acquired through experience.[65] Although these two factions were divided in their thought, it would have been possible for them to overlook their differences and pursue the truth with the mixture of their ideas. René Descartes, who was a rationalist, stated that the human mind is equipped with pure reason that contains, within itself, certain truths that are experienced independently. John Locke, an English empiricist, argued that humans are born with a blank slate and that their knowledge is constructed progressively from their experiences through their sensory perceptions of their surroundings. It is interesting to notice that this view of the human mind can be seen as early as the writings of the Greek philosopher Aristotle, in his *De anima.* [66]

It is easy for us today to agree with the empiricist way of thought. However, we can certainly find some notions embedded within rationalism that might bring both sides to a complete version of one unified theory. A newborn child is not born with the entire universal knowledge that is forgotten when it exits the womb. Newborns learn how to talk, walk, and move in correspondence to their surroundings. Yet something, perhaps "natural light" as Descartes called it, must drive them to do so. What is this natural light? I believe it is their survival instinct.

This instinct that is an intuitive hard coded behavior encourages the newborn to keep learning and observing its surroundings for it to continue its existence. Therefore the origin of reason, and ultimately thought is our survival instinct.

In the next chapter we will tackle different subjects in orbit around this statement. We will focus our discussion on how something so rudimentary, such as bacteria, evolved throughout millions of years to form complex animals with complex behaviors, such as thought. Along the way, we will try to paint a vivid picture of how survival instinct was not only implemented within every step of these evolutionary processes but was also the sole building block for its instigation.

Chapter VI: The Abstraction of Beasts

I would like to detour before we go on. As we shall see, it would not be such a sharp turn from where we are headed in the next chapters. For years reason, thought, and human development have been inseparable from language. Although it is an important element to our growth as an intelligent race, and probably the next best thing that happened to us since the evolutionary introduction of our complex fingered hands, it is unquestionable that communication between species first evolved to alert humans of danger. Animals with this mutation warned others with their mouths and were preceded by animals that perhaps tapped the ground with their feet or released chemical warnings into the air. It is plausible to conclude that listening methods mutated and evolved before calling sounds. To be more specific, hearing, just like smell and sight, helped animals to compensate when one of the senses were impeded by the surroundings.

For example, at night an animal can hear the footsteps of its predator in complete blindness and take evasive actions. The surroundings can overwhelm an animal's sense of smell and is dependent on the movement of air in which the predator's trademarks fly through, but sound is fast and can travel great distances, which makes it an important addition to the survival apparatus of the animal. It is not until animals began living in group settings that they started communicating with each other.

A tremendous amount of research has been conducted to understand animal callings and communication methods. Old and ineffective analytic processes have held back comparative studies of nonhuman communication systems to the complexity of human communication. However, in recent years, scientists have discovered new methods, including the information theory, a discipline in applied mathematics that examines information over a channel or a storing medium to analyze recorded data, which has helped us to understand communications in animals.[67]

From clicks to pitched melodies, we know that animals communicate through sound and sometimes with complexity that begins to resemble our own. We can

51

apply this research to understand how humans first developed language, but we also have other agendas to this research. For example, research on sperm whales and bottlenose dolphins, sponsored by NASA, not only focuses on how or whether they speak but might help to prepare us should we ever be in a "first contact" situation.[68]

It would be virtually impossible to trace the origins of language back to one pre-human group that talked in one common tongue. Although we have a well-documented history of printed text, we have nothing to denote how old spoken language is. It is possible, however, that language came long before its written form. Although purely speculation, it is also possible that tribes were communicating at the time of the earliest cave paintings, circa 35,000 years ago. Often describing a hunt or perhaps a religion or ceremony, these cave paintings were a form of writing.[69] (Comparing the age toddlers start painting what they see and what age they start to communicate through language might give a clue to an early pre-human behavior.)

Some scientists suggest that the origin of speech can be found through paleontological evidence, through studying the vocal areas of pre-human fossils. For example, the Neanderthals lacked the shape of their throat that allowed Homo sapiens to produce complicated vowels. This discovery led to a widely held opinion that Neanderthals lacked language. However, it is possible that they could have had a language that did not require the full spectrum of vowels that we have today. This language could have possibly been quite expressive without the full range of vowels, as the evolutionary psychologist Steve Pinker has remarked.[70]

Looking at modern language and records of old patterns of speech, we could deduce that language developed independently by immigrated tribes throughout the globe. The English language can be traced to Eastern Europe, whereas other languages appeared in other habitats that had no related origins to the English language.

With that in mind, it is important to emphasize that although language did not initiate thought, it is inseparable to human thinking. We talk to ourselves when we solve problems, conceptualizing our own thoughts; yet it would be utterly wrong to assume that animals that do not possess language "abstract not," as was argued by John Lock and others.

Science looked to other animals to solve questions about our own species. Because we developed from lesser species, observing nonhuman animal behavior gives us a unique window into our past.

The more complex the behavior of an animal, the further down the survival instinct origin will be on the chain of thought, often hiding subconsciously. This rule can also be inverted: the simpler the animal is, the easier it will be for us to track the survival instinct correlation to its behavior.

With that in mind, we need to confirm that on some levels, the process of what we call conscious thought is being implemented by the animal exercising its behavior. Before we go on, we must address the question, "Do nonhuman animals think?" We cannot learn about the evolution of our own thought process from nonhumans unless we first establish that animals do in fact think.

Nonhuman Thought

The question of whether nonhuman animals are conscious beings has been the subject of much debate for decades. The subject has been discussed and dissected by philosophy, science, religion, and animal rights organizations—all with their own specific agendas. Philosophers and scientists were captivated by the opportunity to compare human beings to other animals. Religion sometimes situated animals within a paradigm according to which humans, created in God's own image, were viewed as superior beings. Even though theories of consciousness were mostly developed to investigate human beings, these same methods proved exceedingly effective when applied to animals. Since the late twentieth century, there have been more pioneering experiments focusing on the cognitive capacities of animals.

Before we continue to discuss nonhuman thought or consciousness, we should first ponder what consciousness is. No single, lucid notion defines it. On the contrary, the word *consciousness* has several meanings, and we can find more than a few distinct concepts to test the consciousness level in nonhumans.

The initial concept of consciousness is probably the most straightforward one: is the creature awake rather than asleep, dead, or in a coma? If the answer is awake, the more convoluted question to follow will be, is it aware? This sense of consciousness is more important than the first. Awareness means an organism is conscious of its environment and has the basic ability to perceive through its senses and respond. It would not be difficult to agree that nonhumans again pass this test.[71]

Although these tests may seem a bit too rudimentary, every day some unfortunate family struggles with an unimaginably difficult decision when a family member is in a persistent vegetable state (minimally conscious state) induced by head trauma. In a persistent vegetative state, the individual maintains the lower functions of the brain, such as respiration and circulation, yet loses all higher functions of the brain. Even with the help of advanced imaging tools, we are unable to concur that the person is in fact conscious.[72] Often a decision to stop treatment is advised only when the extent of damage to the brain is so immense that it is unlikely that there is any consciousness at all. Yet we are still learning new things about our complex brains. In some cases patients spontaneously recover from a minimally conscious state, and, with a recently developed treatment of patients with severe brain injuries called deep brain stimulation, which involves implanting electrodes with millimeter accuracy to specific areas of the brain,[73] we have new hope to bring people with severe brain injuries back to some level of functionality.

Now that we have introduced two simple senses of consciousness, let us continue with more intricate notions that groups have been using to debate nonhuman consciousness.

Access consciousness requires the animal in question to act with rational and controlled behavior in its own environment. The animal must execute learned behaviors in accordance with the situation in which the animal is placed. This notion of consciousness was first introduced in 1995 by Ned Block, a professor of philosophy, psychology, and neural science. Block suggests that neurological evidence supports that some animals have this level of consciousness and some even exercise high-level cognitive processes that require categorization of perceived information, memory recalling, and planning, which are building blocks for reason and thought.[74]

Numerous studies have dealt specifically with how both lab animals and animals in the wild react to their environment and to each other as groups. I could cite a long list of countless experiments to reach a convincing argument, but instead let's get a bit more personal. Many people own pets, such as a dog or a cat. These animals react to their environment, which is the same principle that must be present for them to be exercising access consciousness. These trained pets learn from our behaviors, sometimes noticing or anticipating them even faster than some humans do.

There are some arguments to consider when using pets as an example. Our observations of our own pets will most likely be quite an unscientific attempt on certain accounts. To begin with, we lack research methods; however, the more profound argument is that we tend to project our emotions and persona onto other animals, just as humans tend to easily identify human-like faces in random shapes. Because we cannot truly get into the animal mind and ask animals how they feel or what they think, we can only assume by using our own experiences. However, as valid as this argument might be it should not only be targeted at the observation of pets. This should be applied to any behavioral research done on animals. We tend to assign emotions to anything with a mouth, a nose, and eyes. We categorize everything that passes through our minds, finding micro- and macropatterns. Could all this research be faulty? I highly doubt it. Animals that share the same biology with humans are most likely not so fundamentally different from us.

A second argument can used to discredit pets as test subjects. Research animals should be observed in the wild to understand their behavior fully because we have corrupted the behavior of domesticated animals. However, this could not be further from the truth. Animals in the wild have been accustomed to their environment for thousands of years. Because we have already established that we are descended from them, we can test animals further than any obstacle with which their natural environment presented them. We teach and test our offspring, too. What would happen if we did not expose them to daily challenges, let alone to sciences, art, and culture? Animals show us that they are more capable than simply swimming in groups, hunting for food, or engaging in sexual acts for instinctual reasons. We teach monkeys to play games, count numbers, and speak in hand gestures. We train dolphins to engage with problem solving games with mind-boggling results. To fully understand the intelligence of an animal is to fully examine its potential for solving problems.

Dogs and cats, although not as smart as dolphins or monkeys, process information and understand their environment. Through training, they can provide us with a great example of access consciousness. They have the notion that they did something wrong through the standards of their training. Both cats and dogs will be less likely to resist the torment of a toddler, a behavior that is impressive and complex. Cats and dogs often act controllably as they were trained to, and in that sense they are fully conscious of their environment.

Next is one of the most controversial definitions of consciousness we have discussed so far. The question this definition asks is whether the animal in question is conscious of its own mental state. The *Stanford Encyclopedia of Philosophy* states that "self-consciousness refers to an organism's capacity for second-order representation of the organism's own mental state."[75] The essential phrase here is *second-order representation.* This is highly difficult to prove in animals, and research has so far yielded results mostly in chimpanzees and limited groups of other great apes.

The controversy stems not only from the definition of nonhuman thought but also as a definition for consciousness at large. Some dispute that self-awareness should not be considered when scrutinizing thought in humans or nonhumans. From a philosophical approach, self-awareness is an appealing notion. However, research in past years shows that from a neurological standpoint there is not much merit to it.[76]

For years our best tool for deciphering whether animals understand their own existence is a manmade flat surface that has been in existence since ancient times—the mirror. Research conducted on monkeys, for example, painted the foreheads of one group with a white mark and the other with no markings. When a mirror was introduced to both the groups, the marked Chimpanzees studied themselves in the mirror and often used their hands to examine the strange addition to their fur, whereas the other group was much less fascinated at their own image.[77]

In recent years, developments in cognitive neuroscience have given us tools to observe, understand, and test many theories concerning the mind. The philosophical notion of self-awareness was put to the test, and, even though this research is still in its infancy, conclusions can still be drawn from the results of these early tests.

Self-consciousness or thoughts of oneself are probably not at all different from the thoughts of others. Research shows that whether a study's object is passive or active after perceiving their surrounding environment, common groups of neurons equally discharge. For example, the study of neurons from area F5 in the premotor cortex of monkeys has shown that they are directly linked to distinct types of motor functions, such as grasping with the hand. Neurons from that group are also subdivided into classes, corresponding to the grip of the object that they involve.[78]

In this process, the test subject will first perceive the object, then, according to the object size, order different actions. You can easily test this for yourself. Place a drinking glass (one with no handle intended for a hot beverage) and place a screw (or any small object) about fifteen to twentye inches from it. Now, pick up the glass, put it down, and pick up the screw. You may have noticed that the grips that were involved with these objects were quite different from each other. One grip was intended for a medium-sized object and involved the thumb and the other fingers opposed to each other. The screw, on the other hand, required a much more precise grip. The precision grip in this case involved two fingers, the index finger and the thumb, opposed to each other. Simple tasks such as these two grips have their own sets of neurons, but, surprisingly, these neurons will discharge regardless of whether we pick up these objects or just look at them.[79]

This might seem a bit off track for our discussion of consciousness, yet important conclusions can be drawn from findings such as these; thought, as complex and as magnificent an evolution achievement as it is, is less mysterious than some would believe it to be. An animal that is aware of its surroundings is constantly processing information from it environment and deciding whether to act. This animal would not only apply its knowledge or experience to others but also to itself.

Earlier we saw how using a simpler life-form, the *Aplysia,* can solve questions concerning a much larger and significantly more complex system. Now that we have eradicated any claims that nonhumans are merely automaton beings, we can consider how they behave and compare our observations to the behavior of our own species.

The Thinking Ancestors

One of the most prized talents humans possess, other than our expressive languages, is our ability to use our hands to create tools. From tools to mega structures, we have pushed the exploitation of our environment to its limits, and we continue to push it further and further from its natural way. Humans have learned to make fire, to use wood, to carve stones, and to build magnificent structures. We have learned to turn different compounds into metals and to construct tools for protection and for daily use. However, the ability to build is not restricted to humans only. Some of our tallest buildings seem miniscule when compared to structures built by termites (in proportion, of course). Insects, whom we considered mindless creatures, build colonies in magnificent detail. From food storage rooms to deep wells for reaching water, to complex tunneling systems for escape and ventilation, and even nurseries, insects are to some degree as capable as humans in designing their environments to allow the continuation of their species.[80] It would be wrong to assume that there are schools and guilds of architecture and civil engineering in which information is passed through generations of insects. In many cases, hard wired automated neural programs are in charge of these functions, yet from observing them we discover that these neural programs could be the origins of our ambition to overcome the elements and limit our exposure to danger by enhancing our environment to suit our needs. Even with these hardwired behaviors, we can notice that both instincts and learning processes are fundamental for the survival of the species.

Although we considered simple creatures, we can also examine larger animal builders. Birds build high, elaborate nests that not only protect their offspring from egg-loving creatures such as lizards but also are structurally strong and insulated from the elements. Another excellent example of bird construction is the nest building of the bowerbirds from the *Ptilonorhynchidae* family. These out-standing birds build extraordinary, complex nests solely to attract their mates.

These bowers are often constructed on a circle of cleared earth with twigs placed almost perfectly in the nest's middle, and at times are designed with highly decorated structures. Not only do they create elaborate structures, they also deco-rate them with flowers and a variety of objects, such as glass, plastic items, berries, stones, and shells. These items are carefully sorted and placed in different groups, each in its specific place on the bower. If items are moved while the bowerbird is away, he will replace it in its original place. This is a striking example of animal ingenuity. The female ability to recognize and fond males with better decorated bowers is also notable. They visit each bower, inspecting the decoration while the male performs his mating ritual, and will end up selecting the most decorated and elaborate bower.[81]

This next step of evolution that is far more complex from the attraction to body fitness is a complex sexual selection behavior that can serve as an example of the origins of the sexual selection of humans (although one might humor and add that in some humans it seems as if this sexual selection never evolved further). We have created multibillion dollar markets with the sole purpose of attracting our mates and exhibiting our sense of possession. From fashion, to home decor, to luxury cars, the list is endless. We use items to show we are better, and in this case more fit as a mate, than the competition. We display our own tastes by selecting which objects are more to our preference, even when it is dictated by the latest advertisement campaign. In owning things, we show our mates that we can pro-vide them with the comforts of life and perhaps even support a future family.

Mammals rarely build structures. Most mammals are not equipped with the right peripherals to allow them to grasp and to place construction materials. Ele-phants, for example, use their trunks to perform relatively delicate tasks. How-ever, they have no need of shelter. Both humans and chimps are born with opposable thumbs. Chimps have no special need for building nesting areas, but they also lack the muscle needed for rotation of the thumb.[82] To creationists, the chimp's missing muscle is another blind shot at "proof" we have not evolved from monkeys. However, our ancestors had to abandon the trees after the immense climate change in Africa and were left exposed to the elements and to predators.[83] They certainly needed upgrades (by no means I refer to intelligently designed upgrades) to allow them to survive. Other than humans and monkeys, only rodents are equipped with hands. The extent to which rodents use their hands is a product of their environment. Squirrels, for example, handle seeds or nuts with precision. They build nests high in trees and hide their food underground. One species of rodents surpasses all of the others. This exemplary animal is the beaver. These mostly nocturnal animals give us an example of awareness and planning in an animal other than a human. They excavate canals, using them as roads to float construction materials to their construction sites. With their limited natural tools, they cut to size and reshape logs to build dams and versatile living areas. Not only do they use materials to build homes but also to adjust water flow of a river to fit their purposes.[84] These magnificent builders practice not only classical conditioning (learn to recognize food, building materials, and water levels, for example) but also operant conditioning (learning to cut logs and use them to their advantage).

Learning from the Environment—Recognizing by Stimuli

The process of learning in lesser animals is heavily dependent on classical conditioning. That is, animals do not entirely memorize an experience; instead, it is mostly built in components that track down positive or negative associations linked with each experience. More advanced life-forms use these simple mechanisms too but will go further by modifying them in probability calculations. The thought process here is complex, but it becomes clearer when broken down into steps.

Consider the example of a human confronting a yellow, round, flat, and glossy object for the first time, we can see that this difference in thought process is not only apparent to more complex animals but also in stages of life of a complex animal:

Toddler: Yellow round shape => put in mouth (taste) => bad taste => withdraws object from mouth

Adult: Yellow round shape => does not put in mouth unless there is a trusted recommendation (candy, perhaps)

Even though the adult has shown fewer steps in this process, many more steps were involved in the conclusion than that of the toddler. With the help of memory, probability reasoning took over, and the result was a rejection to eat the object.

The consequences of performing this experiment to an animal in its infancy are often irreversible. Of course dying from eating a poisoned object is irreversible; however, the irreversible effects to the animal after eating an object that tastes bad with no considerable danger are far more fascinating. The animal will conclude that this object should never be eaten again and will be wary in the future of objects that look similar.

I have purposely given an example of an object that has a specific shape and color. Not only can we place the object in a category of yellow, round objects, we can also easily observe how pattern recognition works to our favor. The processing network can identify a pattern, place it in a category of its own, and use it for later references to draw probable conclusions. Our sensory peripherals have been developed through millions of years of evolution and are equipped with receptors that exist solely for our protection. For example, bad taste is often an indication that something should not be ingested in the first place. This protective mechanism has been fine-tuned through years of evolution to fit the body's chemistry. Taste and smell play an important role in recognizing patterns of harmful scenarios. Together with memory and prewired programs, we can learn about and avoid substances that could harm us or favor other substances that might contribute to our welfare.

Now that we have established a close relationship between nonhuman behavior, such as learning and reacting to the environment, and human, we can examine another intelligent behavior exhibited by humans but also seen in nonhumans. This behavior is important, and it certainly gave us the ability to conquer our world. It was not until people gathered as tribes, clans, and ultimately civilizations that we achieved true supremacy over our environment. Together we hunted animals for food, built great structures, and solved many problems that could not have been deciphered individually. In a later chapter we will bring the human sociality question once more however, we must first institute nonhuman sociality to establish the origins of our own.

Social Intelligence of Lesser Species and the Collective Intelligence

Earlier we discussed communal living of microorganisms, and the cell was an example of an autopoietic system. As we now discuss more complex life-forms, we have to reconcile these microcivilizations with macro ones.

Among vertebrates and insects, sociality is rare. Nearly all mammals, birds, and reptiles live in solitude except when mating or raising their young (of course with exceptions).[85] Few nonhuman species live in highly structured societies, and we shall discuss some of the unique ones. They are not only unique in that they are constructed of structured societies that resemble our own but also in that they have achieved a level of hierarchy that is almost superior to that of humans (although it could be rightfully argued that their success living in massive societies with order and hierarchy that is second to none is the result of their lesser intelligence, or more so the developed sense of self compared to that of humans).

Living in social groups must have a positive response on the survival of an individual creature, because it would have not evolved as a part of its behavior otherwise. Because evolution favors species that survive, we must look at what makes social living a better solution to the survival of some species. Living in social groups presents obvious problems to many animals, including ourselves. First, there is always the risk of contamination of a colony with harmful diseases and parasites. One individual contracting a disease is often enough to facilitate the spreading of it throughout the colony, starting an epidemic that could devastate the entire population. We must not forget, these micro organisms have been too perfected through evolution, and more so, as our awfully distant ancestors, they are hardwired with the same core instinct we are hardwired with. They have been perfected to reproduce, duplicate and ultimately to survive. It is not the case that they have the ambition to eliminate a colony of any animals. They certainly don't have the capacity to understand what it is they are doing as far as we know. It is merely the fact that the sheer amount of individual hosts are allowing them to prosper in this environment.

It is easy to see how a disease can spread when a group is sharing water and food and livinge together in close proximity.

Consider a computer virus. One computer virus can infect virtually all computers connected to the Internet in about forty-eight hours. All this epidemic requires is one hidden program that attaches itself to the operating system then finds where to duplicate itself onto the next one. The virus will hitch a ride on an e-mail or an infected file; they will reach their new host, duplicate themselves, and use that host to reach other computers with which it has been in contact. Soon, hundreds of thousands of computers will be infected until an antivirus update is issued.

Humans and nonhumans alike have seen their share of epidemics throughout the history of their species. Humans have survived through epidemics such as the black plague, which eliminated about half of the population of Europe, reaching an estimated number of 75 million deaths worldwide.[86] Not only were we affected by these epidemics, but we also learned to harness its deadly potential as biological weaponry early in the ancient world.[87]

Nowadays, thanks to modern sciences, we have a far better understanding of how diseases spread, and we have had our share of breakthroughs that have allowed us to exercise our superiority over these microscopic creatures. However, we are still not free of these potential disasters. HIV has shown how resilient a virus can be to modern medicine and is still a pandemic in areas such as Africa. Because these microbiological life-forms have the ability to replicate rapidly, there is always the possibility of a mutated super-strain that could once again rattle the scientific community.

Another apparent shortcoming to living in a social group is competition. Self preservation was never originally intended for the species as a whole. Prehistoric micro-organisms that produced and replicated continued to exist as a species, whereas others that never had that miraculous mistake evidently ceased to exist. So an animal in a community would still have the sense of preservation as an individual programmed by prehistorically hardwired mechanisms.

When millions of individuals live together in a community, there will be competition. The amount of food might not be sufficient to sustain the whole group. Even when food is abundant, there could be a shortage in mates. Since humans evolved into sophisticated cognitive animals, competition is not such a bad idea. The comparison of bowerbirds competing to build bowers to that of human competition to space is a bit of a stretch, yet we can still see the origins from which competition has evolved. We have an ambition to create better, faster, stronger, and deadlier tools than our competitors. This competitive force drove us to shape our civilization as a species. Even though competition can be deconstructive in communal settings, the positives outweigh its negatives. Competition, as in one group competing against another, allows something that would be utterly impossible otherwise: collective effort.

We can certainly observe the advantages of animals hunting in a group in the wild, yet constructing a beehive or sending astronauts to Mars would be unachievable without the collective. Humans have been graced with various individuals who throughout history have made important observations or discoveries as individuals, but they are still using a collective knowledge that was acquired over time, and, in turn, we will be using their discoveries to add to our own. Ants, for example, are not smart animals as individuals; in fact, they are quite inept, yet a colony of ants is impressively clever. As individuals they are incapable of building highways, assigning different tasks, or defending their territory, yet their simple actions as individuals add up to the multifaceted behavior of their collection.[88]

Although there is no chain of multistructured command in ants, at least not one that we are aware of, they are exceptionally organized, each contributing to the success of the group, sometimes to a level enviable by humans. The secret to their superb organization is their simple minds. Humans have tried to implement such communities in their own cultures without success. Nevertheless, it was incorrect to assume we could sustain such a state of communal living. As magnificent as the sagacity of competition of our species is, it also acts as a hindering force to execute this notion of society.

The question remains, how is it that ants achieve such a level of collective intelligence without a remotely sophisticated brain compared to that of higher-functioning animals? Recent studies, such as that of biologist at Stanford University Deborah M. Gordon, have shown that it is probably through smell and touch that ants are capable of such intelligence. The system is fairly simple yet outstanding in its efficiency. In this self-organizing system, ants bump into each other countless times and receive information through smell. For example, an ant that patrolled in the morning would smell like an ant that worked outside. Foragers rely on the rate of interaction or "bumping" with patrol ants to determine whether it is safe enough to go outside and look for food. After receiving enough information that it is safe to go outside, the ant will forage for food and not come back until it is successful. The faster the ant finds food, the faster it will come back; there are no shifts. The same goes for post allocation—one ant can be a nest worker one day, making repairs, and a forager the next day. The collective effort evaluates how many ants will be needed to complete a task, and, if the nest is damaged or if there is a new supply of food, this affects their collective decision.[89] This type of collective behavior was first introduced in the late 1980s. Since then, it has been effectively integrated into various technologies requiring artificial intelligence. Often referred to as swarm intelligence, it has various applications, from army technology, to logistical problems, to Hollywood special effects that simulate massive battle scenes. Perhaps with the advances of nanotechnology, we might be able to fabricate tiny robotic agents to work with collective intelligence to fight diseases, wounds, or cancer within our body.[90]

Another example of swarm intelligence can be seen in more intelligent invertebrates, the bees. Individually, honey bees are smarter than ants. In fact, recent research shows that honey bees are highly sophisticated animals. Individual bees exercise their superb understanding of their environment and the use of shape recognition and memorization processes that are involved in daily decision making. When they work in a hive, bees are exceptional builders. They build structures that can accommodate tens of thousands of their fellow residents. Their natural choice of material is beeswax, a natural wax produced by young worker bees through their abdominal glands. With that wax, the worker ants construct a honeycomb-shaped cell to facilitate honey and pollen storage and to accommodate larvae and pupae. The bee comb is usually built up a branch and is almost entirely exposed, even when the nests are mostly hidden. One particularly interesting behavior that can be observed within some bee colonies that reside in especially hot environments, such as near volcanoes, is that the worker bees mix the wax with stiff and adhesive resin from trees to raise the melting point of the wax, allowing the structure to stay extremely stiff. However, it is not certain whether this behavior was learned by the bees or just a matter of evolutionary trial and error.[91]

Living in a collective body of thousands, communication is essential to coordinate decisions to keep the colony running smoothly. Unlike ants, bees have developed a more complex form of communication, which requires higher cognitive abilities than smell and touch. Their form of communication is far more efficient than touching antennas (one ant can signal another, which will signal to another; whereas one bee can signal many) and faster than using smell stimuli (although they do still use smell). Bees mostly communicate through a form of dance. Through this fashion of messaging, they share a highly sophisticated communication and information system. Using celestial and landmarks to adjust their dancing, they point to food or to other noteworthy interests. Their communication system describes objects or events in time and location.

Juxtaposing bees to ants, we find something surprising in their behavior. Although the beehive is intelligent and uses a form of swarm intelligence like the ants, bees do have the ability to disagree and are encouraged to compete with each other. Bee scouts sent to find a new real estate for a new nest display a fascinating behavior. They check the area for possible nests and return to convince the swarm. Each scout will urge the rest to come and see the location it has found. Their enthusiasm is reflected by the strength of their waggle dance. Together with the directions to the location of the new possible site, the bee tries to convince the rest to come and check the new location. The swarm will then visit each site and will congregate around the agreed suitable site. It is a matter of numbers: fifteen or so bees collected around a site is the threshold of this democratic election of the new nest location. This sort of behavior demonstrates the striking collective mind of the group. Often, the location selected by this voting system is the most suitable for the colony. Not only were they able to solve the solution to a new nest location but also to overcome opinion differences between individuals for the greater good. This is a society we have somewhat managed to copy and use for our own purposes.[92][93]

Yet the true master builders of the insects are the termites. Not only are they the most resourceful species of insects but also the second in line after humans when it comes to construction. Advanced species of termites will build elaborate castles that can rise up to twenty feet. These structures will house royal chambers for the king and queen and ventilation shafts for cooling the inside of the structure and for carrying out dangerous carbon dioxide toxins that accumulate from thousands of inhabitants and from the fungal gardens, which termites use to process their food. Termites are dependent on cellulose. Like many other animals, they do not digest these countless strings of glucose because it is generally indigestible. Termites use two different methods to extract glucose. More primitive termites chew the food together with microorganisms, which will do the work of breaking the bonds that join the sugars in cellulose. These microorganisms are passed through generations by the adults who feed the young on a special liquid that is rich in cultivation. The more advanced species of termites use a far more efficient method. They build gardens of fungi in which the cellulose is digested outside their bodies. Fungi are the only monarchy of organisms capable of digesting and breaking cellulose in the presence of air. Some of the fungus species used by termites can only be found in termite colonies. The termites delicately weed the fungal gardens, treat them with antibiotics to keep bacterial growth to minimum levels, and take them to the open when it is time for the fungi to reproduce.[94]

Even still, nonhuman intelligence was the subject of scrutiny for many years. I would like to believe that we have conveyed sufficient examples to eradicate any doubt. We could have discussed even more examples of the supreme intelligence of dolphins or chimps; however, I wanted to take a different approach. Because our intention is to locate the origins of thought and problem solving, using much simpler animals as our examples seemed more suitable, especially because they provide us with a much older prehistoric window.

The last and probably most attractive human behavior we will discuss is insight. Insight is generally a solution to a problem. The analyst will consider a problem and conceptualize the solution. Humans sometimes describe insight as a sudden feeling of discovery—a eureka moment—but insight also comes through slower processes of probability evaluations. Insight is unique but not exclusive to humans. Although a direct correlation between insight and language has not yet been discovered, language helps us to conceptualize and organize our logical thought with the help of our memory, and, in doing so, insight can be accelerated, which is probably why we are capable of insight more than other animals.

Great works of art and science were considered (and still are in some circles) as gifts from the gods, yet another proof of our heavenly family tree. It is true that we will not see a monkey writing a novel anytime soon. However, insight is not restricted to humans. Although most of the population will consider birds as low intelligence animals, one animal that is almost infamous with ingenuity and insight is the common raven (*Corvus corax*). Although they often receive bad connotations in folklore, ravens have coexisted with humans for thousands of years. These animals are considered the most intelligent species of birds. One very famous experiment devised by Bernd Heinrich sheds new light on the intelligence of this species. In the experiment, a group of ravens in a large cage was presented with pieces of meat that were suspended from a horizontal branch by a string. At first, the ravens attempted to grab the meat midair and to fly off. However, they soon discovered it was a useless and potentially risky effort. Perhaps discouraged, they sat on the branch looking down at the food. The experiment than recorded the most extraordinary behavior. One of the ravens suddenly flew to the branch, pulled the string with its beak, set its foot on the length of string it managed to pull up, and, securing it neatly, pulled more string, and again set its foot on the string it had pulled up. The raven continued with this process until the piece of meat reached it on the branch. The animal did not use trial and error. When it sat on the branch, it had already solved the problem in its mind. Even though this might seem like a simple experiment, it was mind-boggling insight on behalf of the test subjects. The animal had used motorical skills that were at its disposal but rewired them to address a new problem. The raven had to conceptualize the solution, or at least to foresee a possible outcome to future action in its mind. Although we might never know exactly how the raven conceptualized that specific solution in its brain, we can clearly see there was insight. Other ravens have implemented other solutions to get the piece of meat. For example, one grasped the string with its beak and walked sideways while stepping on the retrieved length. This solution was also creative and required the raven to understand the relationship between cause and effect, which is imperative to logical thought.[95]

Several tests with different variations, such as using transparent fishing line or attaching strings to rocks instead of meat, have all proven to be as successful as the first. Even though not all ravens were successful, 100 percent success rates should not be expected. After all, humans have varying levels of success rates in problem solving. The *Corvus* family also showed a remarkable use of tools. The New Caledonian crow strips a twig of its leaves. This leaves the instrument easy to hold in the beak with a sharp harpoonlike end. With this instrument, the crow can retrieve insects from hard to reach areas.[96][97] Yet is this an actual display of ingenuity? Or perhaps it is some lucky coincidence that the raven has copied from others who have passed it through generations.

This question regarding the problem solving capability of crows was solved after further experiments were performed. In one of the tests, a female and a male New Caledonian crow was given two pieces of metallic wire: one was straight and the other had a hook at its end. They were both presented with a transparent tube. At the bottom of the tube was a metal bucket with a piece of meat inside it. The test required the bird to identify the problem, find the tool with the hook to raise the bucket, and eventually retrieve the meat. The male grabbed the hooked wire and ran off with it. The female, however, seized the remaining wire, and, after noticing it would not do in its original form, the crow bent the end of the wire and retrieved the bucket.[98] This clearly shows that the crow is able to solve a problem with effective understanding of artificial objects. Not only did it solve the problem quickly, the crow used a material that is unavailable in its natural environment by clearly understanding the goal and the potential of the material.

Crows continue to provide spate of examples of animal ingenuity. They have been observed cracking nuts by dropping them on a road, allowing large and fast-moving vehicles to do the nut-cracking for them. Some in urban areas have learned to drop the nuts on a crosswalk and wait for the vehicles to stop, when there is far less risk involved.[99] We can never be sure whether their method of nut-cracking was a result of trial and error or an insight. However, the notion that they are acting upon automatic instinct-only would be a myopic oversight. If we were to collect all human inventions that were either a direct or an indirect result of trial and error, we would end up with an extremely large book. In any case, we should not dismiss these as achievements. Even though the method of their discovery was not the work of ingenuity per se, the process of discovering them is nothing short of ingenuity. In trial and error, one must have the insight of what they want to achieve and then use what they have at hand to accomplish the goal. On the other hand, discovering something to be useful requires a much more complex thought process than we usually like to credit. Both processes are not only the product of a complex neural activity but also the same products that allowed humans to become what we are today. Evolution has a long memory, and we can easily abandon our anthropocentric views and observe how we can trace our origins of thought to a much simpler life-form that exists in the evolution of our species and our planet.

That brings us to the end of the discussion on nonhuman thought. We have mentioned the importance of cause and effect to reason. Just like the crow, we picked little by little to get to our sources of information. Now, we can discuss human reasoning and how together with memory they have evolved into human thought.

Chapter VII: The Origins of Thought

Perhaps unconventionally, we will start with the climax of this chapter and then move our way backward. I would like to believe that the answer to the origins of thought should be clear after the preceding chapters. The oldest origin of human thought is our prewired survival instinct. After millions of years of evolution, this instinct was masked under a heap of intelligent behaviors, yet even these can be traced back to the survival instinct.

Our survival instinct brought us this far, and our existence is most likely the purpose of life itself. The thought might be frightening. When one asks the philosophical question, Why are we here? the answer is simple. If we were not here, we would not be asking the question.

When we try to decrypt the human mind and the thought process, we find that thought is rooted in a single foundation. We think for the sole purpose of keeping ourselves alive and well. We use reason to anticipate what will happen if we do or say things. We grow up in societies in which we memorize rules and social norms so we can anticipate the consequences of our actions. This gives us a fair understanding of how to react to our surroundings, which will keep us from harm.

That is not to say that we are continuously and consciously aware of this behavior and are, therefore, cold, calculating robots. In fact, other than life-or-death situations, we pay little or no attention to the survival instinct portrayed in our daily actions and activities.

This entire notion sets aside any dreams of a divine purpose for humans. We were simply lucky to evolve this far, and we could intellectually evolve much farther if only we were able to acknowledge that. To think that humans possess a supernatural gift of divine purpose to life seems quite a conceit and un-pragmatic thought. I do not intend to dismiss the possibility of the existence of some supreme universal force; however, religious groups have stood strong for hundreds of years behind beliefs in a flat earth, demonic possession (also known as epileptic seizure or justification for wrongdoing), and the model of the sun orbiting the earth. They are again shortsighted, looking at their own backyards instead of to the universe.

How is it possible that a straightforward and simple withdrawal reflex automated by a simple neural network, such as the one observed in the *Aplysia*, could ultimately evolve into creative thought? It is not possible. However, combine hundreds of billions of these simple operations, some for motorical and visual objectives, some for the sole purpose of memory in all its various forms, garnish with thousands of years of collective knowledge, and top with years of individual learning, and you will get creative thought. Just as the evolution of the transistor led to supercomputers that can do much more than the original, humans are capable of doing much more than the Aplysia.

We like to discuss great thinkers and inventors when we talk of the accomplishments of humankind. Yet a great deal of processing goes on even when we do our daily activities. From a very young age, we learn how to walk, talk, and differentiate between right or wrong (according to the culture of our surroundings) with the help (or threat) of punishment. We will carry these laws with us and learn much more while interacting and exercising these laws within social groups.

Religion is based upon these societal laws. Dos and don'ts are carved in stone, written on parchment paper, or passed on verbally from generation to generation. Why do we need these rules? My grandmother, who passed away about a year ago, told me a story when I was a child about a man who lost his religion. I cannot seem to remember the full content of the story; however, there was one thing she said I will never forget: no matter of what, it is always good to be afraid of something.

Religion was the best solution for the ultimate punishment of wrongdoing.

A child is frightened from possible punishment from his parents or teachers, just as an adult will be afraid of the hand of the government. Laws can send a man to jail for short periods, life sentences, or to death. However, governing bodies who respected value of human life and social rules were not available to us during the dark ages and much farther back in time. Even though some societies did have elaborate rules and traditions, religion was mixed in to cement their importance.

Some of these religious rules were meant to protect us from others and from ourselves. For example, Jewish traditions, such as the tradition of washing hands before eating or not eating pig, kept sanitation also important and prevented epidemics within the Jewish communities around Europe in the time of the black plague (although these tradition failed to save them from the circus of conspiracy theorists who used their survival statistics against them).

We have enough documentation of laws in the ancient cultures, from the Ten Commandments to the code of Hammurabi, to see how law enforcers dealt with day-to-day regulations that were important to keep a steady state of co-existence in social living. We might ask how this relates to thought.

The answer is simply pattern recognition. The same method our neural networks use to decrypt and encrypt information passed from the senses also allows us to understand our surroundings and to anticipate social behaviors: how one would react to what I am about to do. However complex the reasoning is, there is no real magic here. Pattern recognition developed to this complicated reasoning method to allow individuals to survive. We can see this reasoning in animals and humans, even though humans with their capacity to do so took it to much greater lengths; thus the survival instinct is responsible for all human and nonhuman thought.

It is possible to argue that this whole notion is flawed because humans and nonhumans break the laws of their community, murder, steal, commit suicide, or sacrifice their lives to save others. Not only do they exercise this behavior, but they sometimes do so while recognizing, or not recognizing, the consequences of their actions. This is obviously contradictory to the survival instinct-thought correlation.

In response to this argument I will split my answer to two different parts; one in response to breaking the law and the other to self sacrifice. They are both fundamentally different. We shall start with self sacrifice since this behavior is the most urgent one to this chapter. To address this argument, please allow me to start with the biblical story of Adam and Eve as described in the book of Genesis. In the story, Adam and Eve were presented with two trees in the Garden of Eden. One is the Tree of Life, and the other is the Tree of Knowledge of Good and Evil. God told them to touch neither of these trees. Then something interesting happened. Both naked yet with no sexual awareness ate from the Tree of Kenowledge and became aware of their nudity, that is, their sexuality. Another observation I would like to add is that the word *knowledge* is used in the biblical sense for sexual relations. "To know" is actually to have sexual contact.

There are two options for a species to survive. One is to live for eternity (Tree of Life); the other is to procreate and make copies of the genetic code. Because animals do not live long and living organisms tend to decay and die quite rapidly, the first option is not yet a possibility. Living organisms have survived with a natural selection that is basic at its core. Species that did not procreate simply did not survive. This became apparent from prehistoric germs. So another instinct is presented here: a hardwired instinct that governs us to procreate and survive as a species. It is important to notice that there is a difference between the two types of survival. One is the survival of the individual; the other is the survival of the group.

This brings us to answer the alleged self sacrifice contradiction. Procreating to produce offspring is only one part of the equation. Of course, there would be no point in producing a defenseless creature and leaving it to the wild. Some measures must be taken to guarantee the survival of the offspring. We see in nature all kinds of methods developed by different animals. Some safeguard their young with utmost care until they become individually sustainable. Some produce thousands of offspring and only a few will survive.

Different methods evolved in accordance to the animals' surroundings. Humans most likely produce one of the most vulnerable and dependent creatures and have to go through a detailed process to ensure its safety. Animals not only take care of their own but also others in their collective. As highly evolved and intelligent creatures, humans have developed a sense of care that allows us to be what we define as humane, or humanitarian.

We discussed earlier the creativity of bowerbirds and common ravens. There are numerous examples of other intelligent animals; however, we have yet to discuss the creative mind of a human in detail. Unlike other animals, we are constantly challenging the limits of our knowledge and the arts. However, we tend to neglect that we exercise our behaviors mostly for the same purposes as our nonhuman neighbors. The substantial difference between our behavior and theirs is that we camouflage it through hundreds of thousands of years of evolution. To be more specific, we've even managed to further elude ourselves through a shorter period of only thousands of cultural revolutions.

Just like the bowerbirds, we try to create artifacts that are better than our competitors' are, and we go through a lot of effort to make ourselves preferred candidates with a better standing as a possible mate. Combine the mating struggle with our efforts to survive in our community at large with our incredibly plastic (from plasticity, flexible and reconstructive) brain and you will acquire a creative mind. Unlike the benefits of the creativity of the bowerbirds, humans benefit tremendously from their mating competition. For example, a person could develop an invention to gain riches; something that can reward them with such a fortune must be important or highly required by many others. It is a great tradeoff. It is unquestionable how beneficial competition is for our culture, both in the sciences and arts.

Now, we can go on defending the argument that was presented earlier. Why do humans hurt each other, murder, steal, or commit suicide even though they know the consequences of their actions? The answer can be unfolded to possibilities that are more specific.

The very first approach deals with false reasoning. I do not claim that humans are perfect reasoning machines. On the contrary, some people are, at best, awfully flawed in their reasoning. This could be the result of early education or a mental state which we will discuss in further detail. We all, however, make mistakes in reasoning. Making mistakes is one way of learning what to anticipate in the consequences of our actions.

The second approach is the opposite of the first. The individual knows of the consequences; however, after reasoning, the individual decided that the benefits of the act are greater than the risk. Two extreme cases can be illustrated here. In the first, a hungry homeless man sees an opportunity to steal food from a food stand. The food is left unattended for a few minutes, and the man, who has not eaten for a week, decides to take it, knowing that he might risk losing his hand for stealing. In the second illustration, a highly trained thief considers a high-profile bust from a jewelry store. He knows how to deactivate the alarm and calculate his odds of getting out in case the alarm goes off. The prize is worth about ten million dollars. If he is caught, he could go to jail for some time. However, when he is released, he could retrieve the jewelry from where he hid it and live his life comfortably. There are many examples that deal with various crimes, but the point is the same.

Killing or stealing to gain superiority is a matter of believing in a cause or in its benefit to the community or the state. In that case, humans are capable of being taught or manipulated to perform these kinds of acts. Inquisitions and holy wars have served vast empires for hundreds of years, but both the soldiers and the inquisitionists have had no apparent consequences; some were even considered honorable for dying or killing in the name of a holy cause. This is possibly our worst and greatest achievement in veiling our prehistoric instincts. We are able to either manipulate others to our advantage or to self-destruct against all reason. This brings us to the most influencing element on rational human thought—emotions.

I began my answer to the argument with a biblical reference from the Book of Genesis. My interpretation to the story of Garden of Eden was solely for illustration purposes and I continue to the next section of the answer with another interpretation from the same book, again for illustrative purposes only.

For many years woman have been stereotyped as beings who are more in touch with their emotions. Whether this is true, we can go forward with our biblical reference. After Adam is created in the image of God, a woman is created to keep him company. The literal translation from Hebrew is "to make him a helper as his counterpart," or an even more literal translation, "help against him." Let us entertain the thought that the creation of woman was actually the creation of emotion. The Bible is known for describing things that are not exactly as they appear on the first read. Indeed, there could be no better description of human emotion than the biblical description of the creation of woman.

Emotion and Its Effects on Decision Making

Consider an electric circuit board crammed with numerous buttons, microchips, and flashing colored lightbulbs. Your job is to reverse engineer this device and to map a diagram that fully encompasses the function of each button and microchip. The most obvious way of doing so is to turn the buttons on and off and to examine what happens if you disconnect each of the electrical connections. The other option is to look at hundreds of devices that are constructed the same way as the first but are malfunctioning. The task is certainly daunting. Multiply this circuit board hundreds of thousands of times, and you understand the job of the cognitive neuroscientist. Generally, the more complex a system is, the more things that can go wrong with it. The human brain is no exception. All animals are affected by diseases, viruses, genetic birth deficiencies, and even malnutrition. Humans, on the other hand, have developed such a complex brain that on top of these diseases we can be significantly challenged and handicapped by our own stimuli. More than any other animal, we are sensitive to our surroundings from birth and childhood. This supreme awareness of our surroundings was our only way to survive in the prehistoric ages and is without a doubt vital for many of our attributes. However, it is also our Achilles heel.

The Vulcan race, a fictional race of aliens from Star Trek,[100] is famous for having no emotions. Vulcans are not born without emotions, but they undergo many tribunals to suppress their emotions and thus act solely on rational thought. This is a good example to our own race. The story is not flawed because in its distant past the Vulcan race was even more emotional than the human race. A significant question surfaces from these fictional characters. Would it be possible for intelligent species to become intelligent, let alone survive without emotions? The answer is no. By chance, evolution manufactures the tools that are essential to the survival of the species. Attributes that are unnecessary are dismissed over time (i.e., eyes in some cave dwellers, wisdom teeth in humans) and attributes that are required for survival will be developed by mutation and will prevail by being passed on to the next generation. So why do animals need emotions to survive?

Let us start by looking at nonhuman animals first; in doing so, we will be closer to our distant history, which simplifies the survival instinct-thought correlation.

The first emotion that comes to mind is fear, or a sense of danger. Fear can be subcategorized to knowledge of the known danger or knowledge of the unknown. In both cases the animal will proceed with extra precaution. A sound heard from a distance or the smell of an enemy will alert the animal, and it will either run or hide. But an unknown smell, unknown sound, or unknown surroundings will set the animal on high alert mode as well. Animals that have an underdeveloped sensing of the possible danger would ultimately become extinct.

Another emotion that can be observed in animals is anger, which is a bit more evolved than fear (although down its roots it is again, built upon self survival, yet requires more steps of evaluations of a given situation than fear). Anger is a provoked response to a threat or frustration. In many cases the threat will be of possession, whether materialistic or more abstract, such as a position in a hierarchy.

Happiness is another important emotion in animals. Although happiness is an abstract emotion, it nonetheless originates from and is important to survival. Happiness, like many other systems, is built upon response reflexes, in this case, positive feedback or self-reward.

At its core, happiness is the opposite of a negative feedback response, such as pain. An animal will learn to avoid being hurt after an experience that has produced that feedback. On the other hand, achieving objectives that are important to survival, such as obtaining food and water will satisfy thirst and hunger which are effective alerts for survival, will reward the animal. An animal can be rewarded for many other objectives. Sex, for example, is one of humans' favorite ones, but shelter from danger or the elements is also rewarded with feelings of happiness or comfort.

Although we have not discussed the full spectrum of emotions, let us briefly consider the importance of emotions to human achievement. I could see no better example to amplify the importance of emotions to our achievement than the endeavors we take to protect ourselves from danger. Without fear, we would have not survived, but neither would we have invented tools for protection or built monumental walls and castles. Humans, lacking the needed protective means, such as thick fur, sharp teeth, claws, or supreme vision and muscles, have artificially developed themselves into what they were not given by nature. Equipped with our most valuable tools, our brains, our hands, and our vocal abilities, we have developed methods to exercise these tools and to keep them in shape. This can include listening to music, solving problems, or even engaging in different crafts. When we combine these exercises with our motivation to be better than our competitors for mating purposes, we manage to become a highly creative species.

We mentioned earlier the fear of the unknown. Humans, just like other animals, are afraid of what they do not understand. Fear is a crucial behavior to any species. However, as a highly creative species, we manage not only to find innovative explanations to these unknowns but also to harness them to our own benefit. Folklore explains many phenomena through ideas we can easily understand.

Ghosts, the undead, sea monsters, angels, miracles, and the wrath of the gods were often used to explain earthquakes, volcanic eruptions, and much more. Many of these ideas were later dismissed when scientists and logical thinkers joined the effort to uncover the truths—regrettably, not without a conflict.

Earlier we discussed the different areas of the human brain and their prospective responsibilities as working collective systems. For many years, the study of human emotions was not incorporated in cognitive neuroscience research, because it was considered a vague subject matter. For that reason, the study of emotions was not included in the neurological revolution. However, in recent years, it was suggested that emotions should be an integral part in this field, especially because not only are emotions an integral component to the cognitive process but they are sine qua non to the survival of our species.[101] Given this perspective, we should look more carefully into the physiological processes of emotion and its effect on decision making.

As we discussed in the beginning of this chapter, the most notable survival-related emotion is fear. Because the effectiveness of a danger response is based on the speed of reaction, a mechanism must act as a first responder. Yet there are no real hubs in which specific core functions take place, and more than a few interconnected areas of the brain are required to accomplish these. There are certain areas, however, in which we can notice a direct correlation to certain behaviors. The limbic system, more specifically the amygdala structure, has been shown to play an essential role in initial reaction to fear.[102] By no means is the structure the center of fear; it is active throughout many other reactions to our surroundings. However, studies have shown much evidence that that specific area and its adjacent systems act as first responders in a frightening situation.[103]

As an example, consider a scary moment in a movie. Humans are incredibly sensitive to their surroundings. Film directors, even though they probably never studied neural pathways, understand its power. A dark, silent moment accompanied with quiet music and little sound followed by an intense moment of sharp intimidating sound and a character or element that appears suddenly out of the darkness will most likely levitate the audience an inch off their comfortable theater chairs. This neural pathway is very short; the message is received by initial stimuli, bypasses the cortex, and heads straight to the amygdala. The amygdala then sends a message to the hypothalamus with instructions to activate the pituitary gland, which will produce adrenocorticotrophin hormone, which will stimulate the hippocampus and prefrontal cortex by inducing steroid hormones.

This is only one pathway that can be observed, but longer routes are far more impressive. Information is often sent to the sensory cortex areas for identification and evaluation of the threat. After a threat is recognized, the amygdala is activated and produces emotional responses.[104]

Researchers have asked what happens when the amygdala is removed, artificially stimulated, or damaged through a stroke or a head trauma. In a study of monkeys, it has been shown that the destruction of the amygdala produces cataclysmic impairment of social interactions that are crucial for survival, such as evaluating danger and perceiving it from others' reactions. In humans, however, although some research does concur with the role the amygdala plays in stress induction and in an inclination to pessimism, there has been no evidence for such catastrophic impairment as has been observed in monkeys. Humans may have parallel mechanisms that compensate for this shortfall.[105] [106]

The mechanism of feedback is equally significant to one's survival. A rewards system that animals must have to pursue food, shelter, and sex is also necessary. We are certainly aware of such physiological components. A physiological component such as the neurotransmitter dopamine is a fine example. It is synthesized in the brain inside dopamine neurons from the amino acid precursor L-tyrosine. Humans have found various ways of inducing artificial pleasure that otherwise would have been created moderately and naturally in the brain.[107]

Emotions are both the result of neural activity and chemical activity in the brain. We undergo millions of years of fine-tuning these balances. However, in many cases this balance will be defective and will ultimately result in mental instabilities. This imbalance could be the result of genetically borne illnesses such as retardation, other genetic illnesses such as Parkinson's disease, or even induced after a traumatic event or drug use.[108] [109]

In any case, we will notice features of certain behavioral impairment. Some cause little impairment, such as attention deficit disorder or the common depression (as a tribute to the common cold I refer to this as the common depression because it is very different from clinical depression). It is in my belief that many people are drawn into taking prescribed drugs that were never required before they became somewhat popular in mid-to-high class this decade. The pharmaceutical companies have unquestionably benefited from people reporting their unhappiness or using the word *depressed* a bit too loosely. I have no doubt they are receiving fine treatment from their doctors; however, I do question the necessity of easily prescribed drug treatments.

Other behavioral impairments can be severe. Some humans become aggressive, lose control of their actions or words, or even attempt to commit suicide. Because we are dealing with an exceptionally complex body organ, there is no black or white—only a wide spectrum of deficiencies in which we can find various mental disorders. This brings us back to conclude our answer to the contradiction question. Indeed, human thought has originated from survival; however, that does not mean that human thought is based solely on reasoning. Rather human thought is based upon rationalization, which is a logical justification for a belief that is made through a complex set of mental processes.

We are no strangers to the effects of emotions on our daily rational thought. When we are angered, in love, or even sad, we tend to react in ways we may later regret. However, a mentally balanced human will eventually stabilize, or "come to their senses."

To summarize this section, emotions are not only an integral part of our survival instinct that has evolved since prehistoric times, but are in use daily to aid us in making rational decisions. We will endlessly exercise emotions, reason, to form a model of how we see the world around us, what we like or dislike, or what we might suffer from. What is utterly magnificent in the patternse we construct is that they are highly plastic, and throughout one's life these patters will change, develop, or diminish.

Chapter VIII: The Correlation to the Origins

Ways of Looking at the Past

So far, we have mainly focused our discussion on the behavior of everything but mammals. Now is a suitable time to bring forth evolution's latest creation. It was imperative that we follow the same road the evolution of our species did millions of years ago. After all, observing insects and macrobiotic organisms is just as effective as observing distant stars or celestial events to see the early universe (often referred to as cosmic lookback time). In astronomy, scientists and researchers understand that they can look back in time by looking further in space. This is possible because the time it takes a photon to get to Earth from the time of its event can be calculated in light years to find how long ago the event happened.[110] For example, the latest recorded supernova, SN 2006gy, was vividly observed on Earth for weeks. Its origins were in the Perseus constellation, which is located about 240 million light years away. For a brief moment in time, we had the opportunity to look 240 million years in the past of that region.[111] We can look much further back in time. As it stands now, astronomy is looking back as far as they can, up to the moment of the big bang. Just like any system, its origins should be far less complex then what it is now, giving scientists an understanding of our complex universe and perhaps a unified theory of our universe.

Earth biologists have their own tools for observing the past. We have already discussed the importance of fossils, yet the methods of looking into our past are not restricted to geologic findings. We can learn much by studying genes and tracing our origins through the remnants of an organism. Evolution has a long memory, and, over millions of years of evolution, certain genes remained unchanged whereas others mutated; even their mutated form leaves us with some clues to help us find the origin.[112]

We can compare our own genes to those of other animals to determine our past. This method can reveal the great secrets and write the historical path that we took from our origins. Not coincidentally, in each step of our evolution we can observe a direct influence from the ecological factors in that period.[113] When we discussed the brain, we encountered different observations and various research that was conducted prior to modern-day science. The triune brain theory concluded that the brain was a three-level hierarchical structure: reptilian brain, old mammalian, and the new mammalian brain. This was an early observation that was limited to the scientific findings of its age. However, as restricted as it was, its discovery was not way off the mark. Not only can we trace the geographic origins from which we have descended (a well-researched subject by the Genographic or the Global Gene Project partnership of IBM and National Geographic), but we can find the genetic traces of our ape, reptile, and fish ancestors.

As animals continue to evolve into mammals, so did their brains. Mammals are nature's latest and most sophisticated creatures. Because of the climate changes that occurred during their period, physiological changes had to be made to adapt to new threats. Some changes, such as stronger muscles and supreme agility, were obviously sprouted to help mammals fight for scarce food, whereas others changes, such as fur, were developed in response to the dropping temperatures of their habitat. But the most striking changes were the additions that would later give rise to humans, especially additions to the brain.[114]

Because we have already discussed the importance of the newly developed neocortex, there is no need to reintroduce its value to the mammals. However, we now have to consider how its development introduced a new set of calculating mechanisms that will later be described as logic and thought. Although these new calculating brains are capable of engaging in amazing tasks, there was nothing revolutionary about the new abilities that mammals were beginning to be capable of. Everything was built upon older, less evolved behaviors, but the most significant change was in memory accessibility and memory storage capacity. Mammals were able to learn much faster than their predecessors could, and they were capable of remembering it for much longer periods. The *Aplysia californica* had to undergo a considerable amount of shocks until it learned. Most insects, fish, and reptiles are also quite slow in learning. This, however, was starting to change with the birds and the mammals. Our definition of intelligence should focus on two factors: an animal's ability to learn and an animal's ability to execute what it has learned when it is being presented with a problem. Both are interchangeable, because there is no use of being able to learn something if you are not capable of using the acquired information. This is put to the test in survival contexts. Because most mammals produce few offspring, more factors must be assessed for the offspring to survive. Parents must be more aware and protective of its offspring, and a much more complicated behavior of risk assessment is needed when standing in the face of danger. Animals that were more "intellectually" refined had an advantage over the ones who did not and eventually died out. These refined animals opened the door to virtually all that we are today.

From Instinct to Thought—Implementation in Our Daily Routine Behaviors

Mr. and Mrs. Harrison woke up on an especially cold Monday morning in January. They were both feeling unrest after the long weekend. After committing to a social evening on Friday night with friends at a local restaurant that Mr. Harrison did not really like, with Mrs. Harrison's friends whom he did not really like, Mr. Harrison had a bad start to his three days of supposed rest from work. Mrs. Harrison had an undisclosed awareness that her husband was not too fond of some of her friends, but she thought his way of putting up with it and keeping it to himself was quite charming. Mrs. Harrison did not really like some of her husband's friends either, and, although she was quite vocal about it, she and Mr. Harrison still had a close relationship.

On the way back from the restaurant, they were both very quiet; the drive home was about twenty minutes long. The car passed by a local nightclub they had often visited during their undergraduate years. Although they could not see through the club's tinted windows, the flashes of the lights that managed to pass through were enough to remind them of the place that was now filled with another generation of young university and college goers. Out of curiosity, they both gazed at the sedan's GPS, which had a big, bright clock in it. The clock indicated it was ten past eleven. They both let out short bursts of air from their nose, almost a quiet laugh. The time reminded them of how not too long ago that time indicated their unlikely chances of getting into that club without seeking the help of Jason Pelham, the promotion manager whom they both knew quite well. Not only was he their mutual acquaintance, it was at his study group for the international politics' midterm where they both met. This was not why they laughed though. A few years ago they would have been just beginning their night, but now they had to rush home so they would not have to pay Lisa, the neighbor's daughter, extra money for babysitting late. It was not that they were short in money, but Lisa, a soon-to-be entrepreneur after attending a private school for business academics, had found her profitable market and was charging them at obscure rates. She was not too much of a social butterfly to say the least, and that made them comfortable in leaving her in their considerably big house with their daughter and son.

As they made their way home, a giant sign on the side of the road showed a woman holding a bottle of a shampoo. Her hair was bright and shiny, and she looked overjoyed. The lettering read, "beautiful all day long" and were intelligently designed to be smooth and inviting, almost as if they were forged by milk and butter. Mrs. Harrison gazed at the sign for a moment and gave it little to no thought. She has been using this brand for a year and was reasonably satisfied with the product. A few moments had passed when she shouted out, "Cookies!" It did not take much for Mr. Harrison to know exactly what she meant. They had forgotten to purchase cookies for tomorrow's theater and arts class for their daughter, Helen. They were in charge of the cookies, while other parents were in charge of other things, such as beverages or other refreshments. They could not forget the cookies. If they did, they knew they would be the conversation piece among the other parents, and it might reflect poorly on their kids. Mr. Harrison made a turn onto the next street, and they made their way to the local twenty-four-hour grocery superstore.

After Helen's class, they got together for lunch with three or four other families. Helen and her friends went off to play in one of the rooms, and the parents were left to themselves with some of the younger kids. They both thought the conversations were less than engaging. Mr. and Mrs. Harrison titled these people the "cardboard society"—something which made them both giggle when they glanced at each other. After a few hours of glancing at their watches, they separated their daughter from the rest of the clan, but not before a considerable amount of rebellious gestures from her. They knew that by leaving they would set precedence to allow the other families to leave, who were quietly thankful for their sense of timing. None really wanted to be there, and even the host family would not mind a quiet Saturday evening after the considerable amount of cleaning they would have to do when the others left.

On Sunday the in-laws were scheduled for a visit. The house had to be as clean as if they were arriving with white gloves and performing a military inspection. Grocery shopping and cooking had to be completed before they arrived, and both Mr. and Mrs. Harrison divided their tasks to complete the preparations in time. Mr. Harrison did not understand why the house had to be spotless; after all, it was just their parents and not the pope who was coming, but when he mentioned his thoughts, quite a few less than civilized remarks were exchanged between the two.

Monday morning, as they were getting ready for work and making sure the kids were getting dressed and eating their breakfast, Mrs. Harrison slowly went over the list of the things she had to do at work that day. This did not interfere but was actually a part of her morning routine. Mr. Harrison was doing the same in his mind, although with much less enthusiasm. His research work was not exactly what he had hoped for going to university. They both drove in separate cars to work and, after dropping the kids off at school, they reflected upon their weekend, which did not allow them any rest for the upcoming workweek or the following weekend, when they already had plans.

No, there was no print mix-up in which pages from a boring story found their way into this book. There was nothing unique in the story. It is highly possible that some of us have experienced some of these same things.

I wanted to bring this story into our discussion especially because there was nothing exciting about it. It is merely a depiction of a daily routine, but many examples can be drawn from it when we discuss thought processes, associations, and origins of thought in animal instincts.

Let's dwell on a few of the behaviors that might be drawn from the Harrisons' events. First, we should discuss memory association. There were a few examples in the Harrisons' events of how the brain worked. The cookies, for example, seem to have come to Mrs. Harrison mind unexpectedly. But, of course, some catalyst must have triggered the thought. We discussed how memory and pattern recognition work early in the book, and this is no exception. The road sign had lettering that seemed in her mind to have been forged out of butter or milk. This is enough stimuli to activate many other memories that include direct or indirect association with these two food products. It could be that the "cookies and milk" ritual that humans are used to since childhood set up a pretty good foundation for Mrs. Harrison to remember the cookies but there could be many other association memories that preceded that, even though they were most probably rendered by one of the two or both together. Mr. Harrison did not need a further reminder to understand what his wife meant by shouting out "Cookies!" He was given a sharp reminder of the importance of buying cookies for his daughter's event. Because he had probably already discussed purchasing cookies with Mrs. Harrison, the alarming tone of her voice probably made it easy for his brain to locate the memory and to know exactly what Mrs. Harrison was referring to. From childhood we learn how to read other people. From recognizing facial expressions to recognizing tone of voice in others, we set a fair understanding that guides us as we grow up. Although when we are very young it will mostly be the simple answer to knowing between good or bad, the further we continue to develop the more it will be essential for us to understand how to read these two shades and also many more in order to leave in a social environment.

The tone Mrs. Harrison had used was almost as crucial as the word she had chosen. Mr. Harrison had had many years to practice the interpretation of that tone, which he must have heard numerous times but could also reproduce himself. Small nuances in tone, even a vocalized intake of air, indicated a moment of discovery and urgency. The clock was another example of memory association. The subtle movement of the head and the sudden glance at the clock must have been enough to cause the other person to do the same. Because they were both familiar with the night club they passed, it provided rich memories. An individual can zoom by many words and stored pictures that sprout from a single thought.

One question remains. Where does the survival instinct fit into the Harrisons' events? Mr. and Mrs. Harrison operate both as individuals and as part of a system. They must both work to provide income to their family, and both have to keep their young from harm's way. As a couple, they must learn to behave and respond to each other in a way so living would not become unbearable. This obvious trait was developed long before we were humans. Living in a somewhat structured society was often as important to survival as being a strong individual, and our ancestors who learned to live together and work together did just that. Although they might not agree with each other at times, it was crucial that they act in a subtle way or not at all upon their emotion. That in mind, a couple who has been together long enough would know the set of rules they had created during their period of getting to know each other just as well as the set of standards of behaviors that society taught them. They continue to build further sets of instructions around each other, which could be the result of trial and error, assumption, or even direct conversation.

There are many example of self-preservation in the Harrisons' weekend, yet there are even more examples of preservation of the young. I chose to tell the story of a family specifically for that reason. If you recall, when we first discussed the origin of human thought in the survival instinct, we stumbled upon some possible contradictions to that argument. One was dedicated to the protection of the young. We have brought two options to survival: immortality and procreation. Because humans do not produce too many siblings and especially because we have yet to discover how to stop the biological clock without causing other damage, there is a strong need for our race to be protective of its young. Although this might have been introduced later in evolution, it is directly correlated to the instinct of survival. The Harrisons will keep making decisions throughout their life that might impair their own conveniences.

Getting Closer to the Instinct—The Hot Water Test

I have conducted a fairly simple but effective experiment with a to-go cup from a local coffee franchise. The cup, which contains cold tea water, also has a lid on top and a teabag label sticking out. The test subjects were mostly university students aged nineteen to twenty-five. During the test period, I made sure to sip my tea on several occasions (some accompanied with a universal sigh of drinking a hot beverage). After answering random questions, I asked them to bring their hands toward me (palms facing down) and made the notion that I was about to pour some of my tea onto their hands. All but a few individuals who knew me well quickly withdrew their hands, some with a look of distrust and some accompanied by a loud noise with the withdrawal of their hand. The speed of the withdrawal was mostly dictated by their hand-eye coordination, and often the males who played video games where the fastest to react. The few who knew me well either dwelled a bit longer than the others in deciding to protect their skin or did not move their hands considerably more than its position. In the next experiments, I reduced certain elements one-by-one. I conducted some tests without a tea bag label sticking out, some without the sipping and vocalization noises, and some without both indicators. The results were predictable. The more I reduced the indicators, the more the test subjects trusted that the contents of the cup were harmless.

How was that reaction so different but relative to that of the gill and siphon withdrawal reflex of the *Aplysia*? The experiment was not set to learn how many times I would have to spill hot tea on the test subjects' hands to avoid this from reoccurring. In that sense, the two tests were considerably dissimilar. However, there was no need for spilling hot tea repeatedly, because our test subjects were already predicting what might happen next and took action to protect themselves. They had already learned what might happen by observing and recalling memory. In that sense the two tests show how one reaction is simply an evolved reaction of the other. The test subjects had observed the universally known cardboard cup that contains hot beverages. Because tea is usually drank hot, the tea bag was another indicator that the beverage I was holding in my hand was not at room temperature and safe for contact. To top it all, people who are consuming hot beverages often sip, which is our way of cooling down a hot drink with our lips instead of letting it harm the much more sensitive throat. The test subjects must have used logical reasoning to conclude that the drink was hot and they must avoid spilling it onto their sensitive side of their hands. These indicators were enough to calculate what might happen next, and a signal from the brain to withdraw was issued.

Even though this experiment is simple and focused, it effectively demonstrates the origins of reasoning and, evidently, thought. The test subjects used memory to conduct scenarios in their heads, scenarios that may or may not have happened to them before. They quickly produced a way to protect themselves. We can find the relationship between any simple or complex behaviors to the survival instinct. At times the relationship will be very close, thus palpable, and at times it will be veiled under many layers of thought and memory.

The ability to use our extravagant, plastic memory is the greatest gift in our evolution. However, this gift is entirely rooted in our prehistoric ancestors and will always be the driving force, if uninterrupted, behind every action we take.

Future of the Machines

In the beginning of the book we introduced a working robot named ADM7. This robot was capable of sensing its environment. It was able to see, hear, smell, and sense heat. Even though ADM7 was capable in its awareness to its environment, it was not capable in doing much of anything else. We would have to program it to get from point A to point B. Imagine that about eighty years have gone by since we first built ADM7. The research was successful. Fifty years from now, the pioneering work of our time will seem so far away. On the other side of things, technologies have not only used the ongoing neuroscience research but have also completely changed the way we store information on a computer. Perhaps even a technological breakthrough came about and made it possible for the minicomputer to work with quantum processor. So with blinding speeds and the ability to store unimaginable information, we have built a computer that is—in perspective—what our modern computer today is to a vacuum tube calculator.

Software will not be left behind. Today's research of adaptive learning programs was merely a springboard for the software in the market of 2087. Because of our ambition to create user friendly software, voice recognition commands will be developed for operating systems that not only know the person sitting in front of them but also learn every habit of their user and interact with their user. If we observe the research and the technology that are being developed today, we can see that this future is not farfetched. Researchers have already fused neurobiology with technology. Real neural networks are integrated to work with electronic components. Adaptive systems are also being developed today. Even though still in its infancy, robots, such as the ones developed by Electronics and Telecommunications Research Institute (ETRI), are capable of learning, reading, and displaying emotions.

I hope that the question to whether we can artificially create emotion in a machine was already answered before. Based on simple reward feedback we can give software the ability to pursue what best for its purpose and in opposite to terminate or avoid pursuit of something that is not efficient or even harmful. Although the artificial intelligence of the 2080's will be much different than that of today, we can still see how modern computers can solve problems and learn how to solve them even better than they were programmed initially.

ADM7's name is short for adaptive dynamic machine. Seven was the number of the version. By 2087, ADM's name will be changed to adaptive dynamic affection machine, or ADAM. ADAM is equipped with the latest technology. It can walk by itself to wherever it needs to go. It can engage in a conversation, recognize its owners and all of their family and friends. Thanks to its superb motorical abilities and outstanding recognition power, it can do many things around the house and outdoors. ADAM does not have to learn how to walk when it came out of the factory. ADAM's memory is already equipped with many possible scenarios that come with the factory settings; however, it is capable of learning its environment and bending these memories when it has to perform in a new area.

This sounds like a science fiction story. ADAM is indeed only a fictional robot; however, it is only a matter of time until we make it a reality.

There are many questions left. We do not have to deal with them right now, but our grandchildren or our great grandchildren will. Is ADAM a living creature? What are the rights of a robot?

One of the greatest science fiction writers, Isaac Asimov, composed a set of rules for robots. Asimov, a scientist and a visionary, has already envisioned in his Foundation books the possibilities of such robots. He comprised a famous set of rules for robots called the Three Laws of Robotics:

1. A robot may not injure a human being or, through inaction, allow a human being to come to harm.

2. A robot must obey orders given to it by human beings except where such orders would conflict with the First Law.

3. A robot must protect its own existence as long as such protection does not conflict with the First or Second Law.

These rules are unified in one thing: they exist to protect us humans. Asimov understood the complications that could arise if machines do not follow these rules. What would happen if only one rule applied to these machines? What would happen if this rule were: "A robot must protect its own existence"?

We cannot guarantee that these robots will not develop this in time. After all, there will be mistakes in copying code in robots after billions of copies, especially if an adaptive system is copying them. The same thing happened in nature and made it possible for us to be here.

Could our ambition to create better technology lead to our own demise? Some computer scientists are working today on software that learns how to make itself better. Computer viruses have been working for years, copying themselves as real viruses do. What would happen if we were to develop robots that build other, better robots? Would they then be considered living creatures by our own definition? Much earlier in the book, we have discussed the origins of the living cell. We have mentioned that life began from non-life. The future holds a possibility that we will create machines that are constructed from autocatalytic materials that are able to self-organize and self-replicate themselves. There is a strong chance that this method, which may prove to be very successful in terms of artificial independency of thought and successful execution of tasks, will result in much confusion and much debate in our society. We must thoroughly examine the consequences of our creations so when we do face this reality we will not be drawn against a future of uncertainty and chaos.

Epilogue

If we were to condense this book to one word, *patterns* would possibly be the right one. This word describes how we perceive the world but also how we understand the way the world works. Anomalies that surround us are merely irregular or unexpected events, waiting to be explored and to be categorized. In most cases, anomalies will follow the same rules or patterns as the rest of the universe. As we have seen so far, the universe created everything in its own image. Celestial object are created from the same materials as biological creatures. Humans have learned how to harness energy just as nature did millions of years before us. We have developed cameras and recording devices, long after nature created creatures that could listen and see. Even though technology is perceived to be our greatest accomplishment over nature, everything we have created is a result of observing and harnessing the power that occurs in nature.

This poses a great epistemological conundrum because of our perspective of the universe. Is it possible that, because we are trained to recognize and look for patterns, we simply solve everything by doing just that? How can we possibly know what is actually happening around us? In that case we really cannot know what the universe is. It poses two possibilities, neither of which is comfortable. One solution is that the universe is exactly as we observe it and exactly as we learn it. How else were we able to recreate and harness some of its elements ourselves? We observe the universe in much better ways than our ancestors did. We use technology they did not have, and question our own knowledge. The second answer is simple. We cannot know if we are merely using our only tool to understand the universe and solving it in the way that is only possible by us. This is another new-age twist to the old-age brain-in-vat skepticism. However, it would do us no good if we dwelled for too long about it. Furthermore, it was already answered that if we don't know anything we could not know that we don't know anything.

There are many unanswered questions regarding the universe. Some questions may not be answered in our lifetimes, and other questions will remain a mystery forever. The evolution of intelligent life-forms, however, is not one of those questions. I would like to believe that we have reached the point where we can explain how our intelligence came about.

We live in exciting times in which scientists are learning much about the brain. What once was impossible is now the forefront of research, which will not only give us better understanding of ourselves and improve our lives, but also give us the ability to harness that power of nature and develop artificial intelligence of our own.

References

1. J. J. Hooker M. E. Collinson, and N. P. 2004. Eocene-Oligocene mammalian faunal turnover in the Hampshire Basin, UK: calibration to the global time scale and the major cooling event. *Journal of the Geological Society* 161.2, 161–172.

2. T. S. Kemp. 2005. The origin and evolution of mammals. Oxford; New York: Oxford University Press.

3. T. D. White, G. Suwa, and B. Asfaw. 1994. *Australopithecus ramidus*, a new species of early hominid from Aramis, Ethiopia. *Nature*, 371: 306–312.

4. Lynne A. Isbell and Truman P. Young. 1996. The evolution of bipedalism in hominids and reduced group size in chimpanzees: alternative responses to decreasing resource availability. *Journal of Human Evolution*, Volume 30, Issue 5: 389–397.

5. Bryan Kolb, Ian Q. Whishaw. 2003. Fundamentals of Human Neuropsychology. New York: Worth Publishers.

6. Bryan Kolb, Ian Q. Whishaw. 2003. Fundamentals of Human Neuropsychology. New York: Worth Publishers.

7. Curd, Patricia, "Anaxagoras", *The Stanford Encyclopedia of Philosophy (Fall 2007 Edition)*, Edward N. Zalta (ed.), URL = <http://plato.stanford.edu/archives/fall2007/entries/anaxagoras/>. Accessed July 12.2007

8. Sloan, Phillip, "Evolution", *The Stanford Encyclopedia of Philosophy (Spring 2008 Edition)*, Edward N. Zalta (ed.), URL = <http://plato.stanford.edu/archives/spr2008/entries/evolution/>. Accessed July 12.2007

9. Ridley M (2000). "The search for LUCA." Natural History Nov. issue, pp. 82-85.

10. Westall Frances. "EVOLUTION: Life on the Early Earth: A Sedimentary View." Science 15 April 2005: 366-367.

11. Chrisantha Fernandoa, and Jonathan Rowea. "Natural selection in chemical evolution." Journal of Theoretical Biology

    Volume 247, Issue 1, 7 July 2007, Pages 152-167.

12. Chrisantha Fernandoa, and Jonathan Rowea. "Natural selection in chemical evolution." Journal of Theoretical Biology

    Volume 247, Issue 1, 7 July 2007, Pages 152-167.

13. Balme, D. M. 1962. "Development of Biology in Aristotle and Theophrastus: Theory of Spontaneous Generation". Phronesis: A journal for Ancient Philosophy 7 (1–2): 91–104.

14. Chyba, Christopher; Sagan, Carl 1992. "Endogenous production, exogenous delivery and impact-shock synthesis of organic molecules: an inventory for the origins of life". Nature 355 (6356): 125–132.

15. Miller S. L., and Urey, H. C. 1959. "Organic Compound Synthesis on the Primitive Earth". Science 130: 245.

16. Miller S. L., and Urey, H. C. 1959. "Organic Compound Synthesis on the Primitive Earth". Science 130: 245.

17. W. Ross Ashby. 1947. "Principles of the Self-Organizing Dynamic System", Journal of General Psychology Vol 37, pp. 125-128.

18. Fromm, Jochen. 2004. The emergence of complexity. Hessen: Kassel University Press

19. Bruce Alberts, Alexander Johnson, Julian Lewis, Martin Raff, Keith Roberts, and Peter Walter. 2002. Molecular Biology of the Cell. New York: Garland.

20. Paul E. Ceruzzi. 2003. A History of Modern Computing, 2nd Edition. Cambridge: The MIT Press.

21. Bruce Alberts, Alexander Johnson, Julian Lewis, Martin Raff, Keith Roberts, and Peter Walter. 2002. Molecular Biology of the Cell. New York: Garland.

22. I. Bernard Cohen, George E. Smith. 2002. The Cambridge Companion to Newton. Cambridge: Cambridge University Press.

23. Bryan Kolb, Ian Q. Whishaw. 2003. Fundamentals of Human Neuropsychology. New York: Worth Publishers.

24. Bryan Kolb, Ian Q. Whishaw. 2003. Fundamentals of Human Neuropsychology. New York: Worth Publishers.

25. Bryan Kolb, Ian Q. Whishaw. 2003. Fundamentals of Human Neuropsychology. New York: Worth Publishers.

26. Edmund T. Rolls, Alessandro Treves. 1998. Neural Networks and Brain Function. Oxford: Oxford University Press.

27. Edmund T. Rolls, Alessandro Treves. 1998. Neural Networks and Brain Function. Oxford: Oxford University Press.

28. R. W. Burns. 2004. Communications: an international history of the formative years. London: Institution of Electrical Engineers.

29. Edmund T. Rolls, Alessandro Treves. 1998. Neural Networks and Brain Function. Oxford: Oxford University Press.

30. Andy Clark. 2004. Natural-Born Cyborgs: Minds, Technologies, and the Future of Human Intelligence. New York: Oxford University Press.

31. Bryan Kolb, Ian Q. Whishaw. 2003. Fundamentals of Human Neuropsychology. New York: Worth Publishers.

32. Bryan Kolb, Ian Q. Whishaw. 2003. Fundamentals of Human Neuropsychology. New York: Worth Publishers.

33. Stanley Finger. 1994. Origins of neuroscience: a history of explorations into brain function. New York: Oxford University Press.

34. Eric R. Kandel. 2006. In search of memory: the emergence of a new science of mind. New York: W. W. Norton & Company.

35. Stanley Finger. 1994. Origins of neuroscience: a history of explorations into brain function. New York: Oxford University Press.

36. Paul D. MacLean. 1990. The triune brain in evolution: role in paleocerebral functions. New York: Plenum Press.

37. Bryan Kolb, Ian Q. Whishaw. 2003. Fundamentals of Human Neuropsychology. New York: Worth Publishers.

38. Norman Doidge. 2007. The brain that changes itself: stories of personal triumph from the frontiers of brain science. New York, NY: Viking.

39. Elisabeth S. Vrba. 1995. Paleoclimate and evolution, with emphasis on human origins. New Haven: Yale University Press.

40. Bryan Kolb, Ian Q. Whishaw. 2003. Fundamentals of Human Neuropsy-chology. New York: Worth Publishers.

41. Bryan Kolb, Ian Q. Whishaw. 2003. Fundamentals of Human Neuropsy-chology. New York: Worth Publishers.

42. Joaquín M. Fuster. 2003. Cortex and mind: unifying cognition. Oxford; New York: Oxford University Press.

43. Steven Rose. 1993. The making of memory: from molecules to mind. New York: Anchor Books/Doubleday.

44. Rusiko Bourtchouladze. 2002. Memories are made of this: the biological building blocks of memory. London: Weidenfeld and Nicolson.

45. Warrick J. Brewer, David Castle, Christos Pantelis, Peter Doherty. 2006. Olfaction and the brain. Cambridge; New York: Cambridge University Press.

46. Rusiko Bourtchouladze. 2002. Memories are made of this: the biological building blocks of memory. London: Weidenfeld and Nicolson.

47. Steven Johnson. 2001. Emergence: the connected lives of ants, brains, cit-ies, and software. New York: Scribner.

48. Steven Johnson. 2001. Emergence: the connected lives of ants, brains, cit-ies, and software. New York: Scribner.

49. Jeff Hawkins, Sandra Blakeslee. 2004. On intelligence. New York: Times Books.

50. Eric R. Kandel. 2006. In search of memory: the emergence of a new sci-ence of mind. New York: W. W. Norton & Company.

51. Eric R. Kandel. 2006. In search of memory: the emergence of a new sci-ence of mind. New York: W. W. Norton & Company.

52. Eric R. Kandel. 2006. In search of memory: the emergence of a new sci-ence of mind. New York: W. W. Norton & Company.

53. Steven Rose. 1993. The making of memory: from molecules to mind. New York: Anchor Books/Doubleday.

54. Eric R. Kandel. 2006. In search of memory: the emergence of a new sci-ence of mind. New York: W. W. Norton & Company.

55. Niels Plath, Ora Ohana, Björn Dammermann, Mick L. Errington, Diet-mar Schmitz, Christina Gross, Xiaosong Mao, Arne Engelsberg, Claudia

Mahlke, Hans Welzl, Ursula Kobalz, Anastasia Stawrakakis, Esperanza Fernandez, Robert Waltereit, Anika Bick-Sander, Eric Therstappen, Sam F. Cooke, Veronique Blanquet, Wolfgang Wurst, Benedikt Salmen, Michael R. Bösl, Hans-Peter Lipp, Seth G. N. Grant, Tim V. P. Bliss, David P. Wolfer, and Dietmar Kuhl. 2006. Arc/Arg3.1 Is Essential for the Consolidation of Synaptic Plasticity and Memories. *Neuron* 52: 437–444.

56. Michael F. and Dan-Eric Nilsson. Animal eyes. Oxford; New York: Oxford University Press, 2002.

57. Thomas J. Bruno, Paris D. N. 2005. Svoronos: CRC Handbook of Fundamental Spectroscopic Correlation Charts. CRC Press.

58. Michael F. and Dan-Eric Nilsson. 2002. Animal eyes. Oxford; New York: Oxford University Press.

59. Michael F. and Dan-Eric Nilsson. 2002. Animal eyes. Oxford; New York: Oxford University Press.

60. Eric R. Kandel. 2006. In search of memory: the emergence of a new science of mind. New York: W. W. Norton & Company.

61. Dale Purves. 2004. Neuroscience, 3rd ed. Sunderland, Mass: Sinauer Associates.

62. Thao Doan, Roger Melvold, Carl Waltenbaugh. 2005. Concise Medical Immunology. Philadelphia: Lippincott Williams & Wilkins.

63. Reason. *The Oxford Dictionary of Sports Science,* Oxford University Press, 1998, 2006, 2007. http://www.askoxford.com/concise_oed/orexxason?view=uk, accessed July 12, 2006.

64. Matthias Steup. Epistemology. *The Stanford Encyclopedia of Philosophy* (Fall 2006 Edition), Edward N. Zalta (ed.), http://plato.stanford.edu/archives/fall2006/entries/epistemology/, accessed July 12, 2006.

65. Jason S. Baehr. A Priori and A Posteriori. The Internet Encyclopedia of Philosophy (2006), http://www.iep.utm.edu/a/apriori.htm, accessed July 12, 2006.

66. Shields, Christopher. Aristotle's Psychology. *The Stanford Encyclopedia of Philosophy* (Winter 2005 edition), Edward N. Zalta (ed.), http://plato.stanford.edu/archives/win2005/entries/aristotle-psychology, accessed July 12,2006.

67. Joshua B Plotkin, Martin A Nowak. Language Evolution and Information Theory. *Journal of Theoretical Biology*, Vol. 205, No. 1. (7 July 2000), pp. 147–159.

68. Jennifer L. Miksis-Olds, John R. Buck, Michael J. Noad, Douglas H. Cato, Dale Stokes. Analysis of Australian humpback whale song using information theory. *The Journal of the Acoustical Society of America*. November 2006, Volume 120, Issue 5, pp. 3228–3229.

69. Thomas Heyd, John Clegg. Aesthetics and rock art. 2005. Aldershot, Hampshire, England; Burlington, VT: Ashgate.

70. Steven Pinker. How the mind works. 1997. New York: Norton.

71. Allen, Colin. Animal Consciousness. *The Stanford Encyclopedia of Philosophy* (Winter 2006 edition), Edward N. Zalta (ed.), http://plato.stanford.edu/archives/win2005/entries/consciousness-animal, accessed March 30, 2006.

72. Chris Borthwick. The Permanent Vegetative State; Ethical Crux, medical Fiction? (1996), http://home.vicnet.net.au/~borth/PVSILM.HTM, accessed March 30, 2006.

73. T. Yamamoto, T. Tsubokawa, Y. Katayama. 1994. Deep Brain Stimulation in a Persistent Vegetative State: Criteria for Selection of Candidates and Long-Term Follow-Up Results. *Stereotactive and Functional Neurosurgery*, Vol. 63, No. 1–4: 63:7–7.

74. Colin Allen. Animal Consciousness. *The Stanford Encyclopedia of Philosophy* (Winter 2006 edition), Edward N. Zalta (ed.), http://plato.stanford.edu/archives/win2005/entries/consciousness-animal, accessed March 30, 2006.

75. Allen, Colin. Animal Consciousness. *The Stanford Encyclopedia of Philosophy* (Winter 2006 edition), Edward N. Zalta (ed.), http://plato.stanford.edu/archives/win2005/entries/consciousness-animal, accessed

76. Colin Allen. Animal Consciousness. *The Stanford Encyclopedia of Philosophy* (Winter 2006 edition), Edward N. Zalta (ed.), http://plato.stanford.edu/archives/win2005/entries/consciousness-animal, accessed March 30, 2006.

77. Gordon G. Gallup Jr. 1970. Chimpanzees: Self-Recognition. *Science*, Vol. 167. no. 3914, pp. 86–87.

78. Akira Murata, Luciano Fadiga, Leonardo Fogassi, Vittorio Gallese, Vassilis Raos, and Giacomo Rizzolatti. 1997. Object Representation in the Ventral Premotor Cortex (Area F5) of the Monkey. *The Journal of Neurophysiology* Vol. 78 No. 4: 2226–2230.

79. Akira Murata, Luciano Fadiga, Leonardo Fogassi, Vittorio Gallese, Vassilis Raos, and Giacomo Rizzolatti. 1997. Object Representation in the Ventral Premotor Cortex (Area F5) of the Monkey. *The Journal of Neurophysiology* Vol. 78 No. 4: 2226–2230.

80. James L. Gould. 2006. Animal architects : building and the evolution of intelligence. New York: Basic Books.

81. James L. Gould. 2006. Animal architects: building and the evolution of intelligence. New York : Basic Books.

82. James L. Gould. 2006. Animal architects : building and the evolution of intelligence. New York : Basic Books.

83. M. E. Hooker and N. P. Collinson. 2004. Eocene–Oligocene mammalian faunal turnover in the Hampshire Basin, UK: calibration to the global time scale and the major cooling event. *Journal of the Geological Society* 161.2, 161–172.

84. James L. Gould. 2006. Animal architects : building and the evolution of intelligence. New York : Basic Books.

85. James L. Gould. 2006. Animal architects : building and the evolution of intelligence. New York : Basic Books.

86. Ole Jørgen Benedictow. 2004. The Black Death, 1346–1353: the complete history. Woodbridge, Suffolk, UK; Rochester, N.Y., USA: Boydell Press.

87. Eric Croddy. 1997. Chemical and biological warfare : an annotated bibliography. Lanham, Md: Scarecrow Press.

88. Eric Bonabeau, Marco Dorigo, Guy Theraulaz. 1999. Swarm intelligence : from natural to artificial systems. New York: Oxford University Press.

89. D. M. Gordon.1999. Ants at Work: how an insect society is organized. Free Press, Simon and Schuster. 2000 paperback, W. W. Norton.

90. Ray Kurzweil. 2005. The Singularity Is Near: When Humans Transcend Biology. New York, New York. Viking Adult.

91. James L. Gould. 2006. Animal architects: building and the evolution of intelligence. New York : Basic Books.

92. James L. Gould. 2006. Animal architects: building and the evolution of intelligence. New York : Basic Books.

93. Eric Bonabeau, Marco Dorigo, Guy Theraulaz. 1999. Swarm intelligence: from natural to artificial systems. New York : Oxford University Press.

94. James L. Gould. 2006. Animal architects: building and the evolution of intelligence. New York : Basic Books.

95. Bernd Heinrich. 1999. Mind of the raven: investigations and adventures with wolf-birds. New York: Cliff Street Books.

96. G. R. Hunt and R. D. Gray. 2004. Direct observations of pandanus-tool manufacture and use by a New Caledonian crow (Corvus moneduloides). *Animal Cognition* 7: 114–120.

97. Ben Kenward, Alex A. S. Weir, Christian Rutz, & Alex Kacelnik. 2005. Behavioural ecology: Tool manufacture by naive juvenile crows. *Nature* 433, 121.

98. Robert Winkler. April 23, 2003. Crow Makes Wire Hook to Get Food. *National Geographic.* http://news.nationalgeographic.com/news/2003/04/0423_030423_crowtools.html, accessed March 30, 2006.

99. John M. Marzluff, Tony Angell. 2005. In the Company of Crows and Ravens. London: Yale University Press.

100. STAR TREK and related marks are trademarks of CBS Studios Inc

101. Richard D. Lane, Lynn Nadel. 2000. Cognitive neuroscience of emotion. New York: Oxford University Press.

102. Richard D. Lane, Lynn Nadel. 2000. Cognitive neuroscience of emotion. New York: Oxford University Press.

103. Richard D. Lane, Lynn Nadel. 2000. Cognitive neuroscience of emotion. New York: Oxford University Press.

104. Richard D. Lane, Lynn Nadel. 2000. Cognitive neuroscience of emotion. New York: Oxford University Press.

105. Bryan Kolb, Ian Q. Whishaw. 2003. Fundamentals of Human Neuropsychology. New York: Worth Publishers.

106. Richard D. Lane, Lynn Nadel. 2000. Cognitive neuroscience of emotion. New York: Oxford University Press.

107. O. Arias-Carrión, E. Pöppel. 2007. Dopamine, learning and reward-seeking behavior. *Act Neurobiol Exp* 67 (4): 481–488.

108. Richard D. Lane, Lynn Nadel. 2000. Cognitive neuroscience of emotion. New York: Oxford University Press.

109. Bryan Kolb, Ian Q. Whishaw. 2003. Fundamentals of Human Neuropsychology. New York: Worth Publishers.

110. Norman K. Glendenning. 2004. After The Beginning: A Cosmic Journey Through Space And Time. London : Imperial College Press.

111. NASA. May 07, 2007. NASA's Chandra Sees Brightest Supernova Ever. NASA online. http://www.nasa.gov/mission_pages/chandra/news/chandra_bright_supernova.html, accessed December 22, 2007.

112. Kenneth M. Weiss, Anne V. Buchanan. 2004. Genetics and the logic of evolution. Hoboken, NJ: Wiley-Liss.

113. Kenneth M. Weiss, Anne V. Buchanan. 2004. Genetics and the logic of evolution. Hoboken, NJ : Wiley-Liss.

114. Nicholas H. Barton. 2007. Evolution. Cold Spring Harbor, NY: Cold Spring Harbor Laboratory Press.

978-0-595-50133-5
0-595-50133-8

www.ingramcontent.com/pod-product-compliance
Lightning Source LLC
Chambersburg PA
CBHW030348290526
45785CB00004B/1657